A CULTURAL HISTORY OF INSECTS

VOLUME 1

A Cultural History of Insects
General Editor: Gene Kritsky

Volume 1
A Cultural History of Insects in Antiquity
Edited by Gene Kritsky

Volume 2
A Cultural History of Insects in the Medieval Age
Edited by Gene Kritsky

Volume 3
A Cultural History of Insects in the Renaissance
Edited by Sherilyn G.F. Smith

Volume 4
A Cultural History of Insects in the Age of Enlightenment
Edited by Matthew S. Lehnert

Volume 5
A Cultural History of Insects in the Age of Industry
Edited by Carol Anelli and Susan Fisher

Volume 6
A Cultural History of Insects in the Modern Age
Edited by Robert K.D. Peterson

A CULTURAL HISTORY OF INSECTS

IN ANTIQUITY

VOLUME 1

Edited by Gene Kritsky

BLOOMSBURY ACADEMIC
LONDON • NEW YORK • OXFORD • NEW DELHI • SYDNEY

BLOOMSBURY ACADEMIC
Bloomsbury Publishing Plc
50 Bedford Square, London, WC1B 3DP, UK
1359 Broadway, New York, NY 10018, USA
29 Earlsfort Terrace, Dublin 2, Ireland

BLOOMSBURY, BLOOMSBURY ACADEMIC and the Diana logo
are trademarks of Bloomsbury Publishing Plc

First published in Great Britain 2024

Series design by Raven Design
Cover image: Detail of ancient Egyptian sarcophagus, XXVI
Dynasty © DEA/G. DAGLI ORTI/Getty Images

A catalogue record for this book is available from the British Library.

Library of Congress Cataloging-in-Publication Data

LCCN 2023029669

ISBN:	HB:	978-1-3500-0305-7
	Pack:	978-1-3500-0321-7
	ePDF:	978-1-3502-6646-9
	eBook:	978-1-3502-6645-2

Series: The Cultural Histories Series

Typeset by Integra Software Services Pvt. Ltd.
Printed and bound in Great Britain

For product safety related questions contact productsafety@bloomsbury.com.

CONTENTS

ILLUSTRATIONS

CONTRIBUTORS

Ian Beavis, PhD, Collections Management Officer, Tunbridge Wells Museum and Art Gallery, UK

Ron Cherry, PhD, Professor Emeritus, University of Florida, USA

Rory Egan, PhD, Professor, University of Manitoba, Canada

Joyce M. Filer, PhD, Researcher, Formerly British Museum, UK

James N. Hogue, PhD, Lecturer, California State University at Northridge, USA

Gene Kritsky, PhD, Dean and Professor, Mount St. Joseph University, USA

Eva Panagiotakopulu, PhD, Senior Lecturer, Professor, University of Edinburgh, UK

SERIES PREFACE

This six-volume set, *The Cultural History of Insects*, is a unique examination of how insects have infested the human experience. The set is divided into six time periods: Antiquity, the Medieval Age, the Renaissance, the Age of Enlightenment, the Age of Industry, and the Modern Age. Within each time period, eight topics are addressed: the state of entomological knowledge during the period; insects and disease; insects and food; insect products; insects' role in mythology and religion; insects as symbols; the portrayal of insects in literature and language; and the representation of insects in art. This structure is meant to allow readers to tailor their reading to their specific interests: for example, if insects' influence in the history of art is of primary interest, the reader can select the art chapter from each volume; if a reader wants to understand the impact of insects on a particular period in human history, then reading the volume on that time period would be appropriate.

The coverage of the topics presented is not encyclopedic. Presented are histories and case studies describing how insects influenced human culture and survival. The geographic focus is generally the Western Hemisphere, but details are included of geographic intersections in which cross-cultural influence plays an important role in various topics and time periods.

Insects were here before humans. They have influenced our evolution, eaten our food, provided us with raw materials, changed our history, and been the muses for our creativity. These six volumes illustrate how insects have helped to shape our culture over the past five thousand years.

Gene Kritsky,
Mount St. Joseph University,
Cincinnati, OH

Introduction

GENE KRITSKY

There has never been a period in our evolutionary history when we were not living with insects. They have fed upon us, eaten our food, slept in our beds, and even played an important role in the development of our culture. Insects first appeared in the fossil record during the Devonian, some four hundred million years ago, long before both the dinosaurs and mammals. Insects were the first animals to fly under their own exertions. Some insects developed their own architecture, farming, language, and spread throughout the globe. They survived the extinction at the end of the Paleozoic and the extinction that wiped out the dinosaurs sixty-five million years ago. Insects are tenacious (Grimaldi and Evans 2005).

We share a common ancestor with the chimpanzees that lived between five to six million years ago. Thus, we can look at our evolutionary cousins' interactions with insects to gain possible insights as to how early humans interacted with insects. The most obvious interaction with insects was as a food source. Chimpanzees eat ants, bees, and termites. All three live in large colonies, which would have provided our common ancestor with a caloric bonanza. Bees, in particular, produce honey, a sweet food that is clearly enjoyed by both chimpanzees and humans. Honey bee stings were a deterrent, but five million years ago, early humans were clothed with a dense covering of hair that today still protects chimpanzees against bee stings (Kritsky and Smith 2018).

Jane Goodall observed chimpanzees selecting long thin stems and inserting them into termite nests. Termites would bite this strange intrusion, which would be carefully withdrawn by the chimp who would greedily eat the insects as if they were candy on a stick. To obtain this treat required the chimps to make a tool, and Goodall's observations documented that humans are not the only toolmakers (Goodall 2022).

Other chimpanzees have been observed taking larger thicker sticks and using them to tear into bees' nests to get the honey. Surprisingly, the chimps carried this tool kit around with them to aid in their quests for liquid gold. These chimps were developing a material culture (Kritsky 2017). This may have happened long before the Australopithecines appeared three million years ago. That opening sequence in Stanley Kubrick's movie *2001: A Space Odyssey* may have gotten it wrong. We may not have initially made tools to kill other hominins. We may have made tools to eat bugs and honey.

Ectoparasites played their part in the evolution of our culture and in our interactions with each other. Genetic studies of the genus *Pediculus*, more commonly known as lice, have revealed that the common ancestor of the chimpanzee, gorilla, and human lice lived some thirteen million years ago. When ancestor of modern gorillas diverged from the chimpanzee/human common ancestor approximately eleven million years ago, their lice diverged as well. When chimpanzees and humans split between five and six million years ago, the chimp and human lice split too. Lice were not just an ectoparasite feeding on us. They played an important role in our socialization. All three of the great primates engage in grooming behavior, where one individual would carefully scour through the hair of another individual looking for lice and their eggs or nits. This is thought to be an important social behavior between infants and mothers and between males and females. This grooming has been and is still taking place in humans, just ask any parent whose child has come home from school with head lice (Kritsky and Smith 2018).

It has been proposed that the loss of our dense covering of hair, which started around 1.2 million years ago with *Homo erectus*, was favored by natural selection to help reduce our ectoparasite load. This loss of our natural insulation led to our wearing clothing. Yes, even fashion may hold its origin to our interaction with insects (Kritsky and Smith 2018).

When we first became conscious of insects is unknown, but there is evidence that we were aware of insects during prehistory. This should not be a surprise to us. Especially when we look at our stone tool industry. The artist Michelangelo allegedly said, "Every block of stone has a statue inside it, and it is the task of the sculptor to discover it" (Parker 2013). The beauty of some of the stone bifacial hand axes from the Paleolithic period were created by a mind that could also see the tool inside the stone (Figure 0.1). The mind that could fashion such stone tools was also painting mammoths, lions, bison, and deer on cave walls in France and Spain. This mind was aware of life in the world, and that included insects.

FIGURE 0.1 The Paleolithic mind could see the tool inside the stone. A bifacial hand ax from France. Photograph by Gene Kritsky.

The Neanderthals migrated out of Africa into Europe around four hundred thousand years ago, although we do not have any material evidence of their interest in insects, it is probable that they shared the same interest in honey as their precursors (Meyer et al. 2016). Our species, *Homo sapiens*, evolved approximately three hundred thousand years ago in Africa with the same knowledge of insects that Neanderthals inherited from our common ancestor (Hublin et al. 2017). It is likely that our interest in insects as food and our sweet tooth for honey continued as *Homo sapiens* migrated into Europe and Asia.

The Late Paleolithic cave paintings and etchings that show woolly mammoths and other extinct animals do have some geometric paintings that some authors have interpreted to be insects, but there have been other interpretations of these markings that are not insect based. The oldest physical artifact that clearly depicts an insect is an incised bone carving of a camel cricket that dates to the Late Magdalenian culture, possibly as early as fourteen thousand years ago. It was found in the rock shelter Enléne in France. Camel crickets, with their jumping hind legs and rounded thorax, often inhabit the entrances of caves and were likely a common sight at the shelter. We do not know what these insects meant to the artist, they might have been a small snack or a target for children to hunt or for play (Kritsky and Smith 2018).

A carving thought to represent the larvae of a bot fly, dating between fourteen thousand and eleven thousand years before the present, also caught the attention of Late Paleolithic artists (Guthrie 2005). A detailed jet carving of a warble fly larva, prominently displayed in the Vienna Museum of Natural History, is tapered like a warble fly larva, and it includes lines that reflect the insect's characteristic segments. There is also a carved warble fly larva on an arrow straightener in London's British Museum that was produced by the Inuit, who prized warble-infested caribou as a delicacy. Perhaps that was true for people living fourteen thousand years ago in Europe.

More direct prehistoric evidence of insect interactions comes from the discovery of pottery shards with the chemical residue of beeswax. Such shards date back to nine thousand years before the present and show that robbing honey bee colonies for their golden sweetener continued well after our species first evolved (Kritsky and Smith 2018).

To find the oldest representation of humans interacting with insects we need to go the Late Neolithic in the eastern Iberian Peninsula to examine Levantine rock art (LRA). These paintings have been suggested as possibly dating to the Late Paleolithic, but there is a growing consensus that they are much younger. The LRA differs from the older prehistoric paintings in that LRA is found in rock shelters, whereas the Paleolithic paintings were produced deeper in caves. Moreover, Paleolithic paintings include extinct animals such as woolly mammoths and woolly rhinos whereas LRA does not. Paleolithic paintings rarely have human figures, unlike LRA which show humans engaged in hunting deer in addition to robbing honey bees' nests. Indeed, it is these honey-hunting LRA paintings that provide us with the most definitive evidence of our interest in insects. There have been eight locations that have been cited as examples of honey hunting, but there are two that are more complete. The best-known example shows two people robbing a wild colony of honey bees. The honey-hunting painting was discovered in 1920 in a rock shelter near Bicorp, a small village approximately 90 kilometers southwest of Valencia, Spain (Figure 0.2). The painting is one of many images painted on the walls of the rock shelter called Cuevas de la Araña, or Caves of the Spider, and it is one of the most widely reproduced images in the history of entomology. It is small, measuring only 55 cm in length. It shows two cross branches at the top to which three long ropes are attached. Near the top of the painting is a small hole, which represents a natural cavity. Even today, honey bees nest in natural cavities in the rock walls of the region. Next to the hole is a thin waisted honey hunter held in place by the rope holding a bag in one hand and presumably harvesting honey from the nest as fifteen bees fly around the honey hunter. Further down on the rope is another honey hunter, who is wearing a sack on his back and climbing

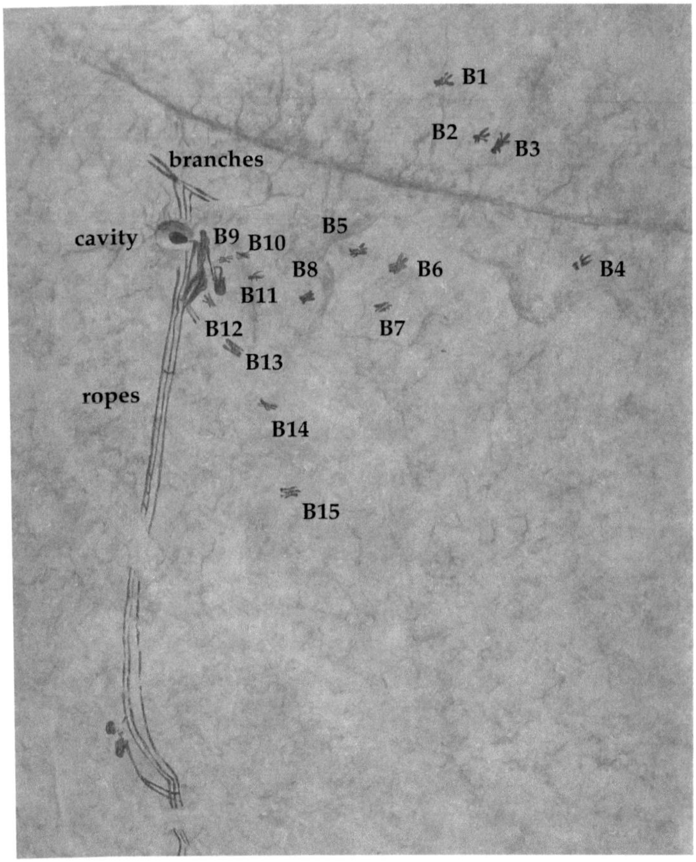

FIGURE 0.2 The honey-hunting scene as painted in 1924 by Francisco Benitez Mellado. The bees are labeled B1–B15. The upper honey hunter is next to the cavity, and the second hunter is located below the break in the lines of the ropes. Modified by G. Kritsky from original in the public domain, http://simurg.bibliotecas.csic.es/viewer/image/CSICAR000145950/1/LOG_0003/.

up the rope. It is now thought that this scene was painted no earlier than 3000 BCE at a time when the hunter gatherers living in the Valencian Neolithic were transitioning from a hunting to a farming way of life (Kritsky 2020).

The development of hieroglyphs in Egypt began around 3100 BCE, and it was not long before insects were used as glyphs in this early form of writing. Crudely carved hieroglyphs of honey bees date to the Second Dynasty. In the British Museum is a small plaque with a crude honey bee that represented the Nile Delta where honey bees were plentiful. The glyph was combined with the glyph of the sedge plant and two half circles to complete the royal titulary that translated to "King of Upper and Lower Egypt" (Kritsky 2015) (Figure 0.3).

FIGURE 0.3 The cartouche of the prenomen of the Eighteenth Dynasty Pharoah Tuthmosis III. The beetle hieroglyph is the triliteral "kepher" and the honey bee hieroglyph representing the Delta region or Lower Egypt. Photograph by Gene Kritsky.

Eventually, Egyptian hieroglyphs included flies, scarab beetles, grasshoppers, preying mantids, and centipedes. Their paintings of nature included butterflies that were accurate to the point where the species could be identified by the painting. Rings, bracelets, and amulets of beetles, flies, and butterflies continued this entomological fashion into the Common Era (Kritsky 1993).

This insect fashion continued through Egypt's Late Ptolemaic Era and into the art of ancient Greece and Rome. Carved gems, called intaglios, occasionally included insects such as beetles, flies, butterflies, wasps, and even stylized scorpionflies (Kritsky and Smith 2018).

Insects' greatest impact during the age of antiquity involved their role spreading diseases. Malaria has been a major disease throughout human history. It is the result from the infection of a protozoan pathogen (*Plasmodium*) and is transmitted by the bite of a female mosquito belonging to the genus, *Anopholes*. Genetic studies suggest that the ancestral malarial genus *Plasmodium* has been in existence for over a hundred million years. The *Plasmodium falciparum*, one of the malarial species that infect humans, split eight to twelve million years ago from the Chimpanzee *Plasmodium*, just before the time humans and chimpanzees diverged from each other. *Plasmodium falciparum* moved into the Fertile Crescent as humans migrated northward. Thus, malaria was present in ancient Egypt and Mesopotamia at the start of antiquity (Hume et al. 2003; Moss et al. 2017).

Although the ancient Egyptians had specific words for honey bee and scarab beetles, both insects that had an association with gods, they did not have a unique word specifically for mosquito, or even malaria for that matter. The Egyptians were certainly aware of biting insects. The hieroglyphic word "*hmy, hnws* and *hnms*" was apparently used in the way we use the word bug. Its use suggests that it was intended to describe a bothersome insect such as mosquito, flea, or gnat (Hansen 2001; Dickson 2006). This vocabulary limitation should not be a surprise to us. Even the old English word for mosquito was "gnaet" (OED Online 2021). To determine if the Egyptians were aware of the disease requires that we examine the archeological record to determine if suitable mosquito breeding habitats were present and to search Egyptian medical papyri to determine if the ancient physicians were treating patients for the symptoms of malaria.

Herodotus (2013) reported in his *Account of Egypt* that the Egyptians were afflicted by biting insects to the point that they developed protection in the form of an insect net:

> Against the [mosquitoes or gnats], which are very abundant, they have contrived as follows: – those who dwell above the fenland are helped by the towers, to which they ascend when they go to rest; for the [mosquitoes or gnats] by reason of the winds are not able to fly up high: but those who dwell in the fenland have contrived another way instead of towers, and this it is: – every man of them has got a casting net, with which by day he catches fish, but in the night he uses it for this purpose, that is to say he puts the casting-net round about the bed in which he sleeps, and then creeps in under it and goes to sleep: and the [mosquitoes or gnats], if he sleeps rolled up in a garment or a linen sheet, bite through these, but through the net they do not even attempt to bite.

The Egyptians use of mosquito netting extends back to the Fifth Dynasty, and the bed found in the tomb of Queen Hetepheres was constructed to be used with netting (Hansen 2001).

In addition to the use of nets to physically keep biting insects away, the ancient Egyptians also made repellents. The Ebers Papyrus, a medical text from ancient Egypt, included a concoction that would be rubbed on the body to discourage insects from biting. It reads, "to prevent *hmnx* from biting: fresh moringa oil. Smear [oneself] with [it]." Another recipe, used to repel biting flies, involved smearing the fat of the *genu*-bird over the body (Ziskind 2009). Herodotus reported that the Egyptians also used castor oil as a repellent and grew it intentionally "on the banks of rivers and lakes" (Gaudet 2014).

Compounding the impact of malaria in Egypt was the annual inundation of the River Nile coupled with the irrigation of gardens and fields. The Nile's waters created life as Erdman (1971) summarized from a tomb at El Kab:

> The inundation over, the Nile withdraws, leaving pools of water standing here and there on the fields. This is a busy time of the year for the Egyptian farmer, "the fields are out" and he must "work industriously" so as to make good the use of the blessing brought by the Nile. He can do this better in the

> sultry heat, which during the summer had oppressed both him and his cattle, has at length given way. "A beautiful day, it is cool, and the oxen draw well; the sky is according to our desire," say the people who till the ground, and they set to work with good will, "for the Nile has been very high," and wise men already foretell that "it is to be a beautiful year, free from want and rich in all herbs," a year in which there will be a good harvest ....

The pools of water left behind as the Nile receded attracted mosquitoes to lay their eggs. In the hot Egyptian sun, eggs would have produced bloodthirsty adults in five days with adults emerging daily as time progressed. If the pools dried up, the mosquitoes would have found ample breeding sites in standing water that would have been supplied daily by the irrigation trenches that were continually supplied by water with the use of shadufs, an irrigation tool comprised of a long pole with a bucket at one end and a counterbalance weight at the other. It enabled the farmer to more easily draw water from the Nile or canal and deposit it into the needed trough. The shaduf was introduced by the Hyksos, who invaded Egypt at the end of the Middle Kingdom initiating the Second Intermediate Period (Bunson 1991). The shaduf, which is still in use today, increased the amount of land in cultivation by 10 percent, which in turn increased the spread of mosquitoes.

The Egyptians loved gardens as evidenced by numerous tomb illustrations. They were experienced horticulturists who imported trees from other lands and grew vegetables in the shade of the trees. These gardens were maintained by a system of canals that brought in the Nile's water for irrigation (Figure 0.4).

FIGURE 0.4 Funeral Ritual in a Garden from the Tomb of Minnacht. Facsimile (1921) of original made 1479–1425 BCE. Metropolitan Museum of Art.

The hot Egyptian climate would lead to increased evaporation of the water, which the Egyptians dealt with by creating lakes. In the Papyrus Harris, the Pharaoh Ramses III is recorded to have announced: "I dug a lake before [my temple Medinet Habu], flooded it with [water], planted with trees and vegetation like the Delta" (Torpey 2016). These lakes were planted with papyrus, for the goddess Hathor, and lotus, for the sun god Re. These lakes along with other water features such as smaller basins that were constructed in front of small chapels provided additional breeding sites for mosquitoes.

By the time of the Ptolemaic rule of Egypt, 300 BCE to 30 BCE, it was realized that mosquitoes were most likely to take their blood meal after sunset. There is a hieroglyphic text at Dendera Temple, which was built by the Ptolemies, that tells people not to leave their homes at sunset for a few weeks after the inundation of the Nile when pools of standing water would have been most numerous (Scheidel 2001).

With all the ideal mosquito habitats available in ancient Egypt, there is little evidence that they recognized malaria. Indeed, there has been debate in the past if malaria was present in ancient Egypt (Estes 1993). The Ebers Papyrus, a twenty-foot-long scroll, includes over eight hundred concoctions of more than seven hundred drugs to treat a variety of ailments. One illness, in particular, is characterized by fever and an enlarged spleen, which may be malaria (Moss et al. 2017). The Hearst Papyrus includes several remedies for treating heat, or what more likely were considered fevers (Leake 1952).

Malaria, unlike other diseases, does not alter the appearance of the body, so past examinations of mummified remains remained inconclusive as to the role that malaria may have played in ancient Egypt. However, recent molecular advancements have documented that malaria was prevalent in ancient Egypt. Miller et al. (1994) used an antibody immunoassay to determine if mummified remains dating from 3200 BCE to 700 BCE had traces of the malaria causal agent. They found malaria present in all seven mummies that they tested, providing considerable support that malaria was present in Egypt throughout pharaonic times. In 2010, Hawass et al. (2010) used genetic tests that determined that the Pharaoh Tutankhamun also suffered from malaria. They also determined that relatives of Tutankhamun as well as an unrelated mummy also had malaria, and all may have died from the disease.

The early death of Tutankhamun from malaria illustrates how insect-borne diseases can influence our history. Tutankhamun died young without leaving any heirs, which resulted in succession crisis. His grandfather, Ay, became pharaoh by marrying Tutankhamun's widow, Ankhesenamun. Ay ruled for only four years and was followed by Horemheb, who previously had served as Great Commander of the Army and King's Deputy to Tutankhamun. Horemheb, like his two predecessors, had no heirs. In order to maintain stability in the country, Horemheb appointed Ramesses I as his successor, which initiated the Nineteenth Dynasty (Clayton 1994). Had Tutankhamun lived, he might have produced an heir and the history of Egypt would have been different. We would not have had the great pharaoh Ramesses II. Insects clearly influenced Egyptian history at the height of the Egyptian Empire.

Like the Egyptians, the ancient Greeks did not differentiate between mosquitoes and gnats. They were much more aware of the disease, but they did not associate it with the insects. Homer states that the appearance of the star Sirius (spring and autumn) brings "much fever upon miserable mortals" (Jones 1909).

Xenophon wrote of the death of Agesipolis from a summer fever:

> While [Agesipolis] was engaged in these operations, at midsummer a burning fever seized him. And since he had previously seen the sanctuary of Dionysus at Aphytis, a longing took possession of him at this time for its shady resting-places and its clear, cool waters. He was therefore carried thither, still living, but, nevertheless, on the seventh day from the time when he fell sick, he came to his end outside the sanctuary.

Xenophon (1918a) was also tutored by his father on how to keep his soldiers healthy:

> "Yes, my son, but just as there are menders of torn garments, so also these physicians whom you mention heal us when we fall sick. But your responsibility for health will be a larger one than that: you must see to it that your army does not get sick at all."
>
> "And pray what course shall I take, father," said he, "that I may be able to accomplish that?"
>
> "In the first place, if you are going to stay for some time in the same neighborhood, you must not neglect to find a sanitary location for your camp; and with proper attention you cannot fail in this. For people are continually talking about unhealthful localities and localities that are healthful; and you may find clear witnesses to either in the physique and complexion of the inhabitants; and in the second place, it is not enough to have regard to the localities only but tell me what means you adopt to keep well yourself."

A clue to what conditions were unhealthy was clearly known to Xenophon, who was aware of what had happened to the Athenians during the Sicilian Expedition. Thucydides (2013) wrote in his History of the Peloponnesian war:

> In the meantime the Athenian generals consulted upon the disaster which had happened, and upon the general weakness of the army. They saw themselves unsuccessful in their enterprises, and the soldiers disgusted with their stay;

> disease being rife among them owing to its being the sickly season of the year, and *to the marshy and unhealthy nature of the spot* in which they were encamped; and the state of their affairs generally being thought desperate.
>
> (emphasis added)

These unhealthful conditions were clearly the mosquito-infested marshes in a region that was rife with fever.

The Greek physician Hippocrates (*c.* 460 BCE–*c.* 370 BCE), considered by many to be the father of medicine, did not know that mosquitoes spread malaria. He did, however, recognize malarial symptoms, and he differentiated the disease from typhoid. Hippocrates cites fever symptoms over a hundred times in his *De morbis popularibus*. Some of these excerpts are quite descriptive of malaria.

> Sweats; in all cases much moisture over all the body. These complaints in many cases were unattended with fever, and the sufferers were not confined to bed; but in many others there was fever, as I am going to describe. Those who showed all the symptoms mentioned above were consumptives who suffered pain. When autumn came, and during winter, continuous fevers—in some few cases ardent—day fevers, night fevers, semitertians, exact tertians, quartans, irregular fevers. Each of the fevers mentioned found many victims ....
>
> (Jones 1868)

> Now the ardent fevers attacked the fewest persons, and these were less distressed than any of the other sick. There was no bleeding from the nose, except very slight discharges in a few cases, and no delirium. All the other symptoms were slight. The crises of these diseases were quite regular, generally in seventeen days, counting the days of intermission, and I know of no ardent fever proving fatal at this time, nor of any phrenitis. The tertians were more numerous than the ardent fevers and more painful. But all these had four regular periods from the first onset, had complete crises in seven, and in no case relapsed. But the quartans, while in many instances they began at first with quartan periodicity, yet in not a few they became quartan by an abscission from other fevers or illnesses. They were protracted, as quartans usually are, or even more protracted than usual. Many fell victims to quotidians, night fevers, or irregular fevers, and were ill for a long time, either in bed or walking about. In most of these cases the fevers continued during the season of the Pleiades or even until winter. In many patients, especially children, there were convulsions and slight feverishness from the beginning; sometimes, too, convulsions supervened upon fevers. Mostly

> these illnesses were protracted, but not dangerous, except for those who from all other causes were predisposed to die.
>
> (Jones 1868a)

However, just having a fever is not evidence in and of itself that the patient has malaria. Jones (1909), who argued that malaria was present in ancient Greece, wrote, "If care be taken, malaria is by no means a difficult disease to identify. A good test is periodicity." The terms quotidians, tertians, and quatrans used by Hippocrates refers to the periodicity of fevers; daily, every other day, and every third day respectively.

Another characteristic of malaria is the enlarged spleen. Fever and an enlarged spleen occur together with other diseases such as typhoid, but unlike other diseases the malaria spleen remains large for years. Hippocrates noted the enlarged spleens of people who lived in marshy areas, which he attributed to drinking the marsh water. This symptom was likely not caused by the water, but rather living in mosquito-breeding areas (Jones 1909). Hippocrates also noted the sequence of symptoms for malaria, chills, fever, sweats, exacerbation (Cunha 2008).

The impact of malaria on ancient Greece has been hotly debated. The historical writings notwithstanding, a few classical scholars doubted that malaria was present in Greece. The evidence that malaria was present in Egypt during antiquity raises the likelihood that the disease made its way to Greece. One possible source of introduction was the Athenians returning to Attica following the wars against the Persians in Egypt that started around 460 BCE. This scenario involves the Athenian soldiers contracting malaria from mosquitoes during the approximately five years they were in Egypt. Once malaria was present in Greece, its spread to Rome was inevitable.

That Rome suffered from the ravages of malaria is clear from the writings of the time. Galen, a Greek physician who lived in Rome from 162 CE until his death around 200 CE, wrote, "The question is not about the reality of the disease, but about the meaning of the name, in relation to those fevers which are called tertian, semitertian and fevers with a tertian form" (Salleres 2002). Even though the mosquito connection was not appreciated, there was a recognition that swampy areas were dangerous, as the Roman Scholar Varro (1934) wrote two hundred years before Galen: "Precautions must also be taken in the neighborhood of swamps, ... because there are bred certain minute creatures which cannot be seen by the eyes, which float in the air and enter the body through the mouth and nose and there cause serious diseases."

Aulus Cornelius Celsus (25 BCE–50 CE), a Roman encyclopedist, amassed a wealth of medical information in his *De Medicinal* including details that some consider the most accurate descriptions of the malaria fevers (Celsus 1935).

> Now there follows the treatment of fevers, a class of disease which both affects the body as a whole and is exceedingly common. Of fevers, one is quotidian, another tertian, a third quartan. At times certain fevers recur in even longer cycles, but that is seldom. In the former varieties both the diseases and their medicines are of various kinds ... Now quartan fevers have the simpler characteristics. Nearly always they begin with shivering, then heat breaks out, and the fever having ended, there are two days free; this on the fourth day it recurs ... But of tertian fevers there are two classes. The one, beginning and desisting in the same way as a quartan, has merely this distinction, that it affords one day free, and recurs on the third day. The other is far more pernicious ... Quotidian fevers, however, vary and have many forms. For some begin straightaway with a feeling of heat, others of chill, others with shivering. I call it a chill when the extremities become cold, shivering when the whole body shakes. Again, some desist so that complete freedom follows, others so that there is some diminution of the fever. I call it a chill when the extremities become cold, shivering when the whole body shakes. Again, some desist so that complete freedom follows, others so that there is some diminution of the fever, some recur at the same time the day following, some either earlier or later; some take up a day and a night with the paroxysm and the remission, some less, others more; some set up sweating as they remit, others do not; and in some, freedom is arrived at through sweating, in others the body is only made the weaker.

Celsus (1935) recommends giving food and drink to the patient along with detailed descriptions of what to provide on what days. Others had stranger cures in mind, including eating insects. Pliny describes the administration of bed bugs to treat fevers, "As to the other remedial virtues attributed to [bed] bugs, for the cure of vomiting, quartan fevers, and other diseases, although we find recommendations given to swallow them in an egg, some wax, or in a bean." Pliny (1856) followed this prescription with, "I look upon them as utterly unfounded, and not worthy of further notice."

Malaria spread throughout Rome as its citizens proceeded to change the local environment. The increasing population surrounding Rome led to an increased demand for fish. By 200 BCE the demand for fish led to overfishing the marshes which, in turn, reduced the mosquito larvae feeding fish population. Pliny wrote that the price for fish increased dramatically, and that mullet rarely weighed more than two pounds (Hughes 2014).

Rome was built around the Tiber River basin with rich farmlands that supported the growing of olives, grapes, and grains. The subsequent population growth led to deforestation for more agriculture and to supply lumber for construction. This resulted in raising the water table. Pliny's nephew wrote that the water table was near the surface at his Laurentum estate, and that Rome had major floods in 54 BCE, 23 BCE, and 15 CE. This situation provided optimal habitat for the malaria mosquito to breed (O'Sullivan et al. 2008).

The harbor port at Ostia was initially free of mosquitoes due to the high salt content of the water, but localized floods provided pools of standing water for mosquitoes. The malaria epidemic in Rome, just a few miles away, supports the view that malaria could easily disperse into the town. As the harbor silted up, and the coast moved westward, more mosquito habitat was formed, and by late antiquity Ostia was no longer safe from the "Roman fever" (Sallares 2002).

All the historical anecdotes suggest that malaria was rampant in ancient Rome, but as with Egypt, new molecular techniques are confirming what some historians have suggested for years – that malaria was widespread in the region. In areas where malaria is well established, it can result in continued reinfection that can impact the vigor of the population and alter the age structure by impacting children the hardest. This has been confirmed with the discovery of a large infant cemetery in Lugnano in Teverina in Umbria that was uncovered near the ruins of a Roman villa dating to late antiquity approximately 450 CE. Many of the remains were of stillborn infants, which would be expected as malaria can result in miscarriages in pregnant women with some degree of immunity against malaria. Other remains were of newborn infants. DNA samples from one child, between two and three years of age at death, were found to have the molecular signature of the malaria parasite. That direct evidence of malaria, combined with plant evidence that the burials occurred during the warmer summer months when malarial infections are more common, strongly supports that malaria was widespread in Italy (Soren et al. 1995; Sallares 2002).

The implications of the classical writings of Pliny, Herodotus, Galen, and Celsus, combined with the archaeological record, have provided evidence that by the start of the fifth century of the Common Era Rome, and the surrounding regions, were so racked with malaria that it rendered the population unable to defend itself against invaders that brought about the fall of the western empire (Soren et al. 1995; Sallares 2002).

Malaria was not the only fever disease that impacted the ancient world. Between 430 BCE and 426 BCE a plague decimated the city of Athens. Several diseases have been proposed for the plague including tuberculosis, anthrax, measles, plague, and typhoid fever. A mass grave dating to the plague was discovered in 1994 in Athens, providing scientists with another opportunity to use molecular biology to possibly answer this historical question. Dental pulp from three victims was collected from their skeletons because other studies found that DNA from disease causing bacteria had been isolated from dental pulp. The analysis found DNA from the bacterium that produces typhoid fever in the dental pulp (Papagrigorakis et al. 2006, 2008).

The devastation of typhoid during the plague of Athens is well known as Thucydides, the Athens historian, contracted the disease during the plague and recorded his experience in his book *The History of the Peloponnese War* (Thucydides 2013a).

> In the first days of summer the [residents of Sparta] and their allies, with two-thirds of their forces as before, invaded Attica, ... Not many days after their arrival in Attica the plague first began to show itself among the Athenians. It was said that it had broken out in many places previously ... but a pestilence of such extent and mortality was nowhere remembered.
>
> Neither were the physicians at first of any service, ignorant as they were of the proper way to treat it, but they died themselves ..., as they visited the sick most often; nor did any human art succeed any better. Supplications in the temples, divinations, and so forth were found equally futile, till the overwhelming nature of the disaster at last put a stop to them altogether.
>
> It first began, it is said, in the parts of Ethiopia above Egypt, ... Suddenly falling upon Athens, it first attacked the population in Piraeus, ... and afterwards appeared in the upper city, when the deaths became much more frequent.
>
> All speculation as to its origin and its causes, if causes can be found adequate to produce so great a disturbance, I leave to other writers, whether lay or professional; for myself, I shall simply set down its nature, and explain the symptoms by which perhaps it may be recognized by the student, if it should ever break out again. This I can the better do, as I had the disease myself, and watched its operation in the case of others ....
>
> As a rule, however, there was no ostensible cause; but people in good health were all of a sudden attacked by violent heats in the head, and redness and inflammation in the eyes, the inward parts, such as the throat or tongue, becoming bloody and emitting an unnatural and fetid breath.
>
> These symptoms were followed by sneezing and hoarseness, after which the pain soon reached the chest, and produced a hard cough ....
>
> Externally the body was not very hot to the touch, nor pale in its appearance, but reddish, livid, and breaking out into small pustules and ulcers. But internally it burned so that the patient could not bear to have on him clothing or linen even of the very lightest description; or indeed to be otherwise than

> stark naked. What they would have liked best would have been to throw themselves into cold water; as indeed was done by some of the neglected sick, who plunged into the rain-tanks in their agonies of unquenchable thirst; though it made no difference whether they drank little or much.
>
> Besides this, the miserable feeling of not being able to rest or sleep never ceased to torment them. ... But if they passed this stage, and the disease descended further into the bowels, inducing a ... severe diarrhea, brought on a weakness which was generally fatal.

Unlike malaria, typhoid fever is caused by a bacterium that lives in humans and is spread by contact with an infected individual or by contamination from sewage, and a major player in this contamination are house flies. House flies were as ubiquitous during antiquity as they are now. They have been shown to be "mechanical vectors" of a number of diseases, including typhoid fever. These "flying sponges" were so effective spreading typhoid fever, that L.O. Howard proposed the insects be renamed as "typhoid flies" as opposed to house flies (Cirillo 2016).

The plague started during the summer of 430 BCE after the army of Sparta laid siege of the city-state Athens. One can imagine that the confined large population with poor sanitation would provide an ideal environment for typhoid fever to occur. The accumulation of human waste would encourage population growth of house flies which would in turn facilitate the spread of typhoid fever that led to the decimation of the Athenian population, including the popular Athenian leader Pericles in the fall of 429 BCE. The plague influenced the outcome of the Peloponnesian war between Athens and Sparta thus changing the overall balance of power in the region.

The plague of Athens may not have been the only case of typhoid fever playing a major role in human history. Alexander the Great died in Babylon in 323 BCE, and the cause of his demise has been debated for centuries (Figure 0.5). The accounts are contradictory, and everything from poisoning to a range of diseases have been proposed. A paper in the *New England Journal of Medicine* authored by physicians and a historian noted that his symptoms included "chills, sweats, exhaustion and high fever." It was also recorded that Alexander cried out with abdominal pain, which is thought to be critical in making the diagnosis that Alexander died of typhoid fever as it can result in a perforated bowel (Oldach et al. 1998). A study by Cunha (2004) using the accounts in the Royal Diaries diagnosed that Alexander the Great died of malaria or typhoid fever, with the latter being the most likely. The evidence that malaria was the killer of Alexander the Great will be discussed in greater detail in Chapter 3. Insects were the vectors of malaria and typhoid in either case. If anyone should doubt that insects have made an impact on human history, just imagine how

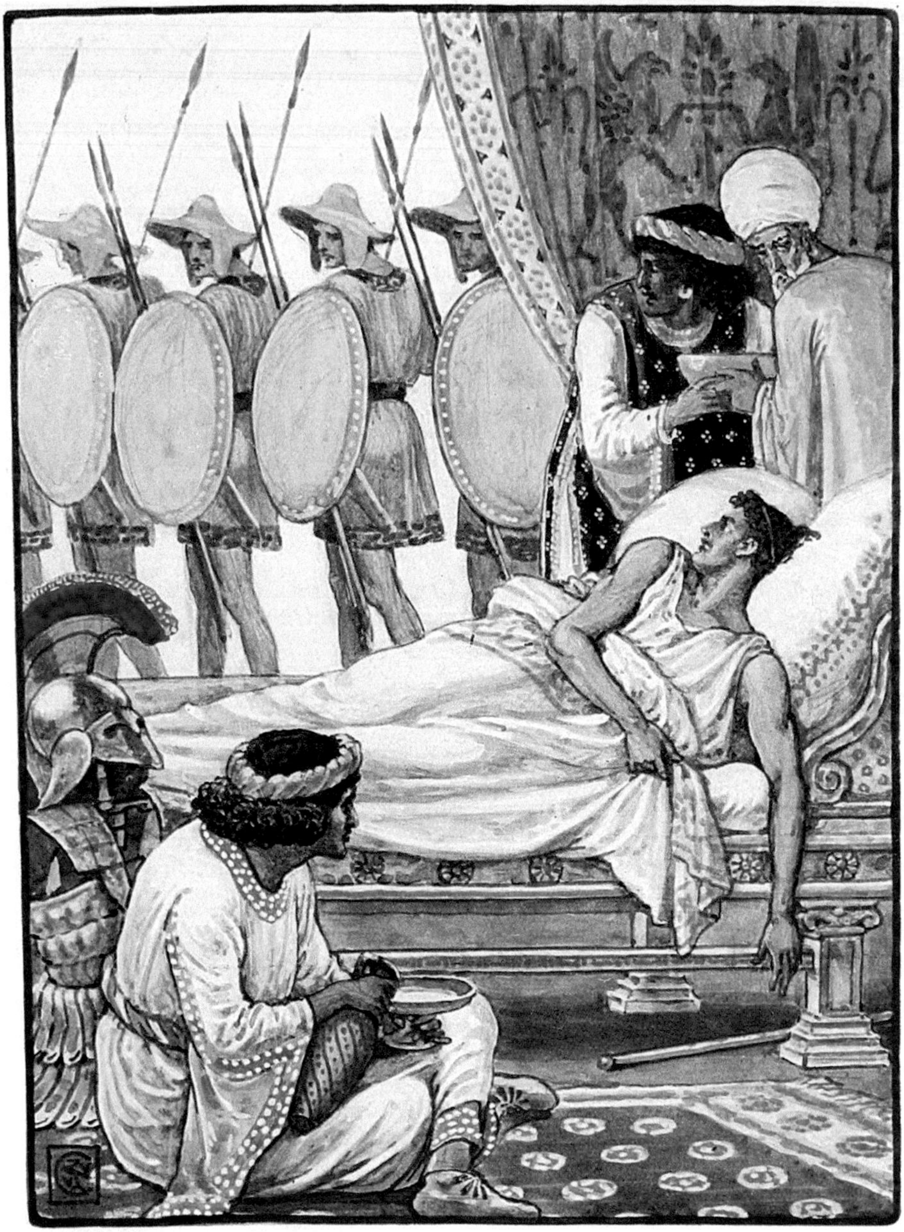

FIGURE 0.5 The death of Alexander the Great by Walter Crane. WikiCommons.

FIGURE 0.6 A bed bug. Photograph by the Centers for Disease Control and Prevention, part of the United States Department of Health and Human Services.

the world's history would have been different had Alexander the Great lived and returned to Greece.

Bed bugs belong to the insect order Hemiptera, or the true bugs, and are characterized by their piercing and sucking mouthparts and wingless condition (Figure 0.6). The origin of the bed bug has attracted considerable attention in recent years because of the nearly epidemic infestations that have occurred with global travel. These infestations likely resulted from hitchhiking bed bugs brought into homes by people, but where did the first bed bug come from? Genetic studies indicate that our bed bugs are older than humans and their ancestor lived long before the rise of mammals, possibly feeding on dinosaurs. If correct, the ancestral bed bugs have been feeding on our distant ancestors back to the Cretaceous (Roth et al. 2019).

The oldest bed bug found in association with humans dates back to ancient Egypt's Eighteenth Dynasty and was found in an excavation of the workman's village Tel el Amarna. The bed bug abdomen was found at a site that was occupied during the reign of Pharaoh Tutankhamun between 1355 BCE and 1346 BCE (Panagiotakopulu et al. 1999).

Bed bugs are mentioned in a 423 BCE play by Aristophanes (1853), a Greek comic and playwright who lived in Athens and Delphi, where his character Strepsaides complains to Socrates about being bitten by bed bugs. Later

Strepsaides says, "I am perishing! The [bed bugs], coming out from the bed, are biting me, and devouring my sides, and drinking up my life-blood, and tearing away my flesh, and digging through my vitals, and will annihilate me." Clearly bed bugs were so common in ancient Greece that they were suitable for comedy.

Bed bugs continued their bothersome behavior during the Roman Era, and they even caught the ire of the disciple John leading him to perform a miracle to rid his lodgings of the biting creatures. The Acts of John of the Apocrypha, verse 60:1 reads:

> Now on the first day we arrived at a deserted inn, and when we were at a loss for a bed for John, we saw a droll matter. There was one bedstead lying somewhere there without coverings, whereon we spread the cloaks which we were wearing, and we prayed him to lie down upon it and rest, while the rest of us all slept upon the floor. But he when he lay down was troubled by the bugs, and as they continued to become yet more troublesome to him, when it was now about the middle of the night, in the hearing of us all he said to them: I say unto you, O bugs, behave yourselves, one and all, and leave your abode for this night and remain quiet in one place, and keep your distance from the servants of God. And as we laughed, and went on talking for some time, John addressed himself to sleep; and we, talking low, gave him no disturbance or, thanks to him we were not disturbed.

Probably the most unusual reference to bed bugs is found in Pliny, where he describes using crushed bed bug as an antidote for snake bites and poisons. "the [bed] bug, for instance, a most filthy insect, and one the very name of which inspires us with loathing, is said to be a neutralizer of the venom of serpents, asps in particular, and to be a preservative against all kinds of poisons" (Pliny 1855).

There are three kinds of lice, mentioned earlier, that can also take up residence on our bodies: the body louse (*Pediculus humanus humanus*), the head louse (*Pediculus humanus capitis*), and the pubic louse (*Phthirus pubis*). The body and head lice are subspecies of each other, whereas the pubic louse belongs to a different genus and has a different evolutionary history. Body and head lice are small, beige to dark tan, wingless insects which are dorsal-ventrally flattened. They have two small black eyes, short antennae, and sucking mouthparts. Their legs are tipped with a single curved claw that bends back to close upon a single spine on their tibia, which permits the body louse to grab onto a hair or fiber of our clothing as it rides around with us.

Lice are wonderfully adapted to live on us. Both male and female lice feed on blood, and they can feed at any time of the day. Females glue up to ten eggs per day on our clothing, and they can hatch in five to ten days. The newly hatched nymphs also feed on blood during the one to two weeks it takes to mature into an adult.

Like bed bugs, lice have been our companions since prehistory. Lice have been found on mummies and on wigs in ancient Egypt. Indeed, the Egyptian fashion of shaving the scalp and wearing wigs may have served as a way to deter the hitchhiking creatures. Herodotus (1920) wrote of Egyptian priests, "Their priests shave the whole body every other day, that no lice or aught else that is foul may infest them in their service of the gods." Head lice have been found on hair combs dating from 200 BCE to 135 CE from Israel (Mumcuoglu et al. 1988).

The Greek traveler Pausanias who lived during the second century of the Common Era documented the presence of lice when he wrote, "When he was in despair, his wife, who had accompanied him from home, among other endearments placed her husband's head between her knees and began to pick out the lice" (Shilleto 1886). Pliny described several protocols to treat lice and their eggs, which are called nits. "Nits are destroyed by using dogs' fat, eating serpents cooked like eels, or else taking their sloughs in drink." As for treating adults, Pliny added, "viper-broth cleanses all parts of the body of lice and removes itching sensations as well upon the surface of the skin." Pliny documents head lice with his entry, "Lice and other foul vermin of the head, are removed by the application of sea-water … Used in the proportion of one part of albumen to two of salt, … destroys lice and other parasitical insects that infest the hair" (Pliny 1856).

The oldest pubic louse was found in Roman Britain and dates to the early second century of the Common Era (Kenward 1999). Thus, all three lice – body, head, and pubic – were present and bothersome in antiquity, but the extent of their distribution is unknown due to the paucity of archaeological finds. However, they were most likely much more widespread as pubic lice eggs have been found on a two-thousand-year-old Chilean mummy. This suggests that the lice have been hitchhiking on human hosts in prehistory as they made it across the Bering Land Bridge in prehistoric times.

The human flea (*Pulex irritans*) is another ectoparasite with a long association with humans. The flea originated in the New World but has been in the Old World for over 5,300 years. They were found on the iceman Ötzi, the frozen man found in the mountainous border region of Italy and Austria (Dickson et al. 2005). Fleas have been collected in ancient Egypt at Tel el Amarna at the same site where the bed bug was recovered (Panagiotakopulu 2001). Fleas were known in ancient Rome as Celsus (1935a) described a protocol on how to remove a flea that entered the ear canal that involved "If a flea has got in[to the ear], a little wool is introduced in which it becomes engaged and so is extracted."

We have had an intimate relationship with insects throughout our biological and cultural history from prehistoric time and into antiquity. In the chapters that follow, we will dive deeper to discover how these six-legged creatures influenced our history, consumed our food, and fed on our bodies at the same time as they inspired our creativity and influenced our thinking.

CHAPTER ONE

Insect Knowledge

IAN BEAVIS

To understand the place of insects in the intellectual world of antiquity we may take as our starting point the rather unexpected statement in the Hebrew Bible that the semi-legendary king Solomon "discoursed of creeping things" (1 Kings 4.33). Within the culture of ancient Israel, Solomon served as the epitome and supposed fountainhead of a wisdom tradition which was shared across various cultures within the ancient Near East and beyond. Although it is highly improbable that there ever was a Solomonic treatise on entomology—whether pseudepigraphical or otherwise—the fact that this topic was included at all as falling within the expected expertise of the archetypal wise man or scholar is significant.

In the passage in 1 Kings, Solomon is credited, besides a more conventional interest in proverbial wisdom and literary culture, with the study of plants, birds, beasts (essentially mammals in our terminology), fish (including most aquatic life), and the category of "creeping things" which would have included reptiles and amphibians as well as all terrestrial invertebrates. The topics mentioned here are far from being the sum of ancient wisdom, which also comprehended cosmological and theological speculation, moral and ethical standards, practical guidance for everyday life, history, geography, medicine, astronomy, meteorology, the lore of various crafts and industries such as agriculture and architecture, and also what we would term magic although in antiquity this was viewed as a set of practical skills for tapping into supernatural forces. Our author's conception of a Solomonic treatise on insects would no doubt have been something akin to a medieval herbal or bestiary, incorporating the contribution of some insects like the honey bee to human wellbeing, their depredations as pests, their uses in medicine, magic, and ritual, a mixture of curious tales from near and far, and the moral lessons humans could learn from the lives and habits—real or imagined—of these creatures. Many references to insect types fitting these categories can be found in the Hebrew Bible, and a similar range of types and interests to humanity is evidenced in ancient Egyptian culture. Surveys of the insects written about and illustrated in ancient Egypt reveal a familiar range of types shared with ancient Israel and the Greco-Roman culture—including pest species like the louse, mosquito, and locust, the domestic honeybee, and the sacred scarab and other beetles of ritual significance (Kritsky 1993; Kenaway and Abdel-Hamid 2015).

The remarkable range of subjects considered as forming part of ancient near eastern wisdom is mirrored in what Greco-Roman culture classed as philosophy. The astonishing versatility—by modern standards—of a scholar like Aristotle, who, in addition to what we might think of as "pure philosophy," could also turn his hand to such diverse subjects as political science, literary criticism, and zoology, is simply a particularly outstanding example of the comprehensive knowledge to which Greco-Roman scholars aspired. Greco-Roman scholars were less conscious than we are of the existence of discrete areas of research, and the relative smallness of the available corpus of knowledge did not necessitate specialization. An educated Greek or Roman could aspire to some grasp of the whole field of human knowledge, and would make connections between diverse fields, ranging across what we see as the boundary between arts and sciences without awareness of its existence. The encyclopedic Roman scholar Pliny declares that the subject of his *Natural History* is nothing less than the whole universe and life itself (Preface 13). Although he sees himself as the first author to tackle this whole area of knowledge, he is confident of his ability to cover it. He may have missed some facts, but his study of two thousand works by one hundred authors is generally sufficient for the task in hand (Preface 17–18). Taking as his model Aristotle, whom he describes as "a man of supreme eminence in every area of learning" (VIII.44), he manages to include many aspects of human culture, history, and geography, alongside a detailed compilation of existing knowledge of the natural world.

Greco-Roman interest in plants and animals, arising from, in our terms, a wide range of perspectives, demonstrates the holistic nature of classical intellectual activity. Often the natural history of any one creature, as described by an author like Pliny or Aelian—who compiles "fascinating facts" about animals for instruction and entertainment—will cover the entire spectrum of interests from "pure and applied science" through creative literature to ethics, religion, and magic, with perhaps an emphasis on one or other pole. The holistic approach of ancient scholarship offered both advantages and limitations. It enabled fruitful cross-fertilization between different disciplines, but it was also inimical to in-depth study and specialization. Perhaps the conviction that it was possible for a single human intellect to comprehend the whole sum of human knowledge—at least in its essentials—led to a disregard of the finer details that were seen as not essential to an understanding of "the nature of universe and all life." This may go some way towards explaining the way in which ancient entomology—as we might conceive it—makes a promising beginning with Aristotle and his followers towards becoming a genuine science but conspicuously fails to develop in the way that it eventually did in post-mediaeval Europe, within a cultural context that was apparently no more auspicious.

Within these all-embracing scholarly traditions of ancient near-eastern wisdom and Greco-Roman philosophy, the knowledge of insects crosses many modern-day intellectual boundaries which were not recognized in antiquity. The most obvious analogue to contemporary entomology lies within the zoological and taxonomic work pioneered by Aristotle and his protégé Theophrastus. But Aristotle drew much of his raw material from a variety of strands of pre-existing wisdom. And later natural philosophers like Pliny did not give the Aristotelian tradition any superior place alongside these other strands as they continued their independent lives through later centuries. Insects feature in creative literature, in the study of agriculture and human and veterinary medicine, in historical and geographical texts, in contexts related to religion and magic, and as a source of moral lessons for human beings to follow. All of these diverse sources were considered—relatively indiscriminately—to contribute to the sum of human knowledge on the lives and significance of these creatures.

The philosopher and pioneering natural historian Aristotle is conventionally seen as the father of zoology, and therefore of entomology as a subdivision of that larger discipline. If that is so, it is curious to observe that although a brave start was made and remarkable advances in knowledge were achieved, these were not built upon until relatively modern times. Aristotle and Theophrastus are the only surviving ancient authors with any reasonable claim to be entomologists in anything like the modern sense, and there is no evidence of lost authors who rivaled them in this respect. The future lay for many centuries with the essentially derivative and repetitive work of learned compilers like Pliny whose approach—incorporating Aristotle's findings but setting them uncritically alongside gatherings from the whole gamut of classical literature—persists up to the threshold of the modern era in the seventeenth-century *Theatre of Insects*, widely regarded as the starting point for entomology as we know it (Mouffet 1658).

Aside from Aristotle and his immediate followers, to speak of entomology in the Greco-Roman world is to impose a modern construct on a diverse series of traditions whose respective interests in animals were not seen as constituting a coherent intellectual discipline. These traditions, which preceded Aristotle, from which he drew much of his raw material, and which carried on unchanged by his effort at synthesis, were essentially pragmatic and human-centered.

Their interest in invertebrates was restricted to their impact on human beings. What they did not need to know—or what they thought was of no practical utility—they did not investigate.

Farmers were interested in insect pests and how to control them, and occasionally in those that could serve as weather signs or which appeared to have some other practical use. The latter category includes creatures like the fig-wasp which the ancients understood as being necessary for the development of the fruit, although the mechanism that we understand as pollination remained a mystery (Beavis 1988: 212–16). The medical tradition was interested in venomous invertebrates and how to treat their effects, in practical uses such as the employment of leeches in blood-letting, and in the use of various insects, worms, myriapods, and arachnids in pharmacology. There is some interest in human parasites and domestic pests and their control, divided between the agricultural and medical traditions. Medicine grades imperceptibly into magic, which in turn overlaps with religion as in the case of the sacred scarab of the Egyptians which was of great interest to Greco-Roman scholars too (Beavis 1988: 162–4). There is a veterinary tradition essentially parallel with human medicine, dealing with parasites of domestic animals, insects harmful by reason of bites or stings or being toxic when eaten, and using similar pharmacological resources. Also of interest were insects producing useful products—in particular honey, with instructions on beekeeping forming a substantial part of the literature on agriculture, but also less well-understood articles like silk and kermes dye.

Most insects and invertebrates described by Greco-Roman authors outside Aristotle will be found to fall under one or other of these special interests. The only exceptions are a few exotic species—such as the fabulous giant gold-digging ant of India (Beavis 1988: 209–11)—which are described by geographers and historians from Herodotus onwards; and a handful of types—such as the glow worm, cricket, cicada, and rose chafer—which were conspicuous enough to acquire a place in folklore, proverbial wisdom, and creative literature. Even these latter usually had some "function" in human life; for example, the Greeks enjoyed the songs of crickets and cicadas and kept them as pets, and children played with the beautiful shiny green rose chafer beetles (Beavis 1988: 76, 101, 167–8). There are several conspicuous insect types which, because they had acquired no definite role in human life, are entirely or virtually missing in classical texts—for example, dragonflies and most other fresh water insects (Beavis 1988: 89)—and ancient near-eastern sources.

Aristotle inaugurated a tradition of writing on the structure, habits, and life history of animals, in which he was never surpassed. In the case of a creature like the bee, which was of economic importance, this coexisted with an "applied science" or craft—in this case apiculture—which had its own instructional literature. Aristotle used practitioners as sources of information, but a tradition of "applied biology" by which "scientists" provided insights to help practitioners did not develop. Beekeepers studied the lives of bees to some extent, but failed to investigate what they did not need to know, showing remarkable lack of curiosity on topics such as reproduction and identity of the three castes. The issue of whether the dominant bee in the hive—seen anthropomorphically as a ruler rather than as a progenitor—was a "king" (as was generally believed) or a "queen" never was resolved and remained debatable as late as the *Theatre of Insects* (Mouffet 1658: 897–8). The Roman writer on agriculture Columella says that the mythological origins of bees, which he nonetheless describes, are not relevant to farmers. Neither are questions over whether they reproduce by copulation or by getting offspring from elsewhere. Inquiring into such questions is the province of those who seek out nature's secrets, being more agreeable to the educated and literate public with ample leisure than to farmers who are busy folk and for whom such knowledge is of no help in their work (*De Re Rustica* IX.2).

Philosophers like Aristotle and learned compilers like Pliny did exhibit such curiosity, and the educated public described by Columella was interested in their speculations, but such writers were reliant on beekeepers for their basic data and failed to adequately investigate the facts for themselves. Pliny claims that Aristomachus of Soli, one of his sources, studied bees for fifty-eight years, and he also tells of an ex-consul at Rome who used hives of lantern horn to observe activity within (*Natural History* XI.19,49), yet accurate and unequivocal information on many points was all too clearly lacking in Pliny's time. Aristotle is the most well informed by modern standards, but even he suffers from being over reliant on second-hand information. He does set out in his *Generation of Animals* (759a) the important principle that facts ascertained by observation must be given precedence over theories, which is in marked contrast to later reliance on written authority. But he does not employ observation himself to resolve disputed questions. Many problems he leaves unresolved, or decides on purely theoretical grounds.

Aristotle marks a significant advance towards a genuine science of entomology in seeking to draw together the data on insects and other animals that was available from the various technical traditions and combining it with some original observations of his own into a single corpus of knowledge. His approach is followed by his pupil Theophrastus—whose botanical works contain much entomology—and probably by some later members of his philosophical school who seem to be responsible for the non-Aristotelian material in our text of the *History of Animals*.

Aristotle, however, would not have seen himself as an entomologist or zoologist. For him, the study of the natural world is a branch of philosophy, and it is always a means to a greater end. His primary concerns are general issues such as the concept of form and the evidence of order, purpose, and rationality in nature, for which animals happen to provide convenient concrete examples. Although he is concerned to justify the study of animals against those who regarded them as beneath contempt, he acknowledges that detailed study of the heavenly bodies would in fact advance philosophy farther; however, since the latter are largely inaccessible to human observation, the philosopher must be satisfied with what is available here on earth. Insects, or any other animals, are not important or worthy of study in themselves, but only insofar as they shed light on the universe as a whole.

Undoubtedly, once involved in the study of animals, Aristotle's natural curiosity caused the subject to take on something of a life of its own in his hands. Not all of the details he includes are explicitly subservient to his grand design. But it is nonetheless true that Aristotle's "zoology" is essentially a by-product of his philosophical interests and remains subordinate to them. He is only inadvertently the "father of zoology," and then only in terms of developments that occurred many centuries after his own time. There is no question of the study of insects for their own sake, let alone any conception of practical applications of that study by which the technical traditions could be raised to a more sophisticated level.

Whatever his intentions, Aristotle and Theophrastus do nonetheless contain in their works a great deal of what we would call entomology, in the areas of taxonomy, physical structure, habits, and life cycles. Much of their most detailed material has clearly been obtained by practitioners of the various technical traditions identified above. Aristotle's accounts in his major works the *History* and *Generation of Animals* of the habits and lifestyle of the domestic honey bee come into this category, as does Theophrastus' detailed description of the development of the fig-wasp. But alongside this second-hand material, there is undoubted evidence of original observations inspired by disinterested curiosity, and the accuracy of many of these can be demonstrated. Aristotle has seen female wolf spiders carrying their egg cases, and spider hunting wasps stocking their nests with their prey (Beavis 1988: 36, 189). He has a good account of the life history of grasshoppers and locusts and of the cabbage white butterfly, where he correctly details the four stages of its metamorphosis (Beavis 1988: 69–70, 125–6). All of these relate to matters that were of no interest to the technical traditions, as farmers saw no practical gain in learning about the habits and life histories of their pests.

Aristotle is remarkable among ancient authors in rejecting anthropomorphic modes of thought. Regarding insects as if they had human emotions and motivations is endemic in both popular and educated belief from the earliest times: Homer has his brave wasps defending their homes and families (Beavis 1988: 193). Aristotle's more austere accounts contrast markedly with the fanciful embellishments of later natural historians such as Pliny, Aelian, and Plutarch, who for example regard spiders as practising a quasi-human craftsmanship, and attribute to ants and bees a fully developed political community with features such as language, government, and warfare (Davies and Kathirithamby 1986: 59–63; Beavis 1988: 39, 204).

Another unique feature of Aristotle's method is his willingness to reject popular beliefs when these could be demonstrated to be false, rather than simply cataloging them as Pliny does. He abandons the universal belief in the spontaneous generation of insects from inanimate matter where it is plain to him that a particular creature in fact reproduces sexually. He knows that cicadas do not stridulate like grasshoppers, and that the shedding of their skin is a part of their development and not an indefinitely repeatable process of regeneration (Beavis 1988: 96, 100).

Aristotle makes some progress in taxonomy, devoting considerable discussion to methods of cataloging and dividing up the animal kingdom, although he never creates a comprehensive system. He divides invertebrates into four categories, of which three are essentially marine and one (*entoma* or segmented creatures) covers all terrestrial invertebrates besides snails, but there are some creatures he fails to place in any category. Within the *entoma* he fails to disentangle earthworms, myriapods, and arachnids from insects proper (*History of Animals* 523b). He recognizes that there are family groupings among the insects, but does not spell these out except in the case of bees and wasps, and of beetles, and it is only the latter group to which he gives a name (*koleoptera*) which we still use today (*History of Animals* 531b). The Linnaean impulse to establish a set of pigeon holes into which potentially any creature can be fitted is simply not present.

Also absent is the Linnaean impulse to give each entity a name by which it can be identified and distinguished, together with criteria by which others can recognize it. Aristotle does recognize the existence of what we would call species, but in practice has no consistent policy in defining them. Much of the time he works with fairly general categories corresponding to the "types" distinguished in popular speech—such as ant, grasshopper, or butterfly—but occasionally he breaks down a "type" into subdivisions with their own individual diagnostic features, which is a slightly closer approximation to our understanding of species. Spiders, for example, are divided into venomous varieties, web spinners, and free-living kinds (Beavis 1988: 34–6).

Curiously enough, physical descriptions of insects, to us the customary way of distinguishing species, are rare in Aristotle and other ancient natural historians. Distinctions are based mostly on habits and lifestyle, for example the *konops* and the *empis* which equate to two forms of small flies in modern parlance. For Aristotle this last pair are similar in appearance and habits, but different in their origins: the one having its early stages in fresh water, the other in sour wine or vinegar. We can recognize here the biting gnats and mosquitoes on the one hand and the vinegar fly on the other, which are actually quite different structurally (Beavis 1988: 230). But Aristotle is unaware of physical differences, or of the fact that the vinegar fly does not bite, suggesting he has not actually given much attention to examining specimens of the insects in question; that is to say, although he undoubtedly watched insects in the field, he probably did not catch, examine, and compare them. This lack of close examination is also evident in his and Theophrastus' handling of the small insect pests with monosyllabic names that were popularly understood in terms of the damage they cause rather than what they look like, and are therefore never associated with their larger relatives. There is no awareness that the *ses* (clothes moth) is a moth or the *kis* (weevil) is a beetle (Beavis 1988: 136–7, 177–8).

One nomenclatural technique Aristotle does use, which is an advance on popular language, is to give insect names from common speech a new more precise sense, which he then uses consistently to demarcate particular types that he wishes to keep distinct. Aristotle prefers this method to inventing new names, although he does apparently coin new terminology occasionally, as in his subdivision of spiders. Thus he provides separate names for the grasshopper and the locust which were thoroughly confused in general usage, and uses two words normally synonymous for types of wasp that he wishes to discuss separately (Beavis 1988: 62–3, 187–9). Similarly, Theophrastus devises a consistent usage for wood-boring pests that is not found in popular speech, establishing distinct terms for the marine shipworm, the larger terrestrial wood boring larvae, and the tiny woodworm or furniture beetle larva (Beavis 1988: 182).

These refinements of nomenclature do not represent a consistent policy, but rather a tool for a particular purpose; that is to say, precision within Aristotle's and Theophrastus' own works. Aristotle often fails to divide when he could, as where he uses the same term for houseflies, similar biting flies, bluebottles, and other members of the modern order Diptera, the two-winged flies (Beavis 1988: 219–22). There is no ambition to create a universally applicable system that others could employ for their own purposes. Often the criteria given for distinguishing the subjects of particular names are actually inadequate to enable the reader to identify them in the real world.

This ad hoc use of nomenclature—precision where required to meet a particular need, combined with vagueness elsewhere—is also typical of the technical traditions on which Aristotle and his followers draw. Farmers did not see any need to classify and distinguish pest species, so all creatures attacking a particular crop tend to be lumped indiscriminately under one name. The agricultural texts are characterized by wild generalizations, as with the perception that all moths are a threat to beehives, and the probable belief that all butterflies resulted from caterpillars on crops. Medical practitioners, however, did feel the need to distinguish different types of venomous invertebrates such as scorpions, which might require different treatments, or different species within the invertebrate types used as drugs. It is, after all, the needs of the medical tradition that explain why botanical nomenclature is much more highly developed than zoological in antiquity: the ancients were not interested in putting names to wild flowers like modern naturalists, but they did want to know how to identify the right herb as a remedy for a particular condition. Pliny (*Natural History* XXV.8) refers to the existence of herbals with color illustrations to enable medicinal plants to be recognized, although he comments on the limitations of the publishing technology of his day, reliant as it was on accurate copying to avoid degradation of the original visual data.

Thus our pharmacological texts distinguish various types within a group of insects comprising in modern terminology domestic cockroaches, crickets, and the tenebrionid beetle *Blaps*, which were thoroughly confused in popular speech, and cataloged different kinds of *kantharides* or blister beetles of varying potency (Beavis 1988: 78–85, 168–70). It is significant that here where accurate identification by the reader is the aim, physical descriptions are employed, involving color and markings, in a way that Aristotle never does, even accompanied by illustrations. The latter have only come down to us in a very corrupt form which renders them little more than decoration, but the originals may well have been more usable, the present form due to a succession of copyists detached from the reality of the actual subjects (Kadar 1978).

The magical tradition also had reasons to distinguish particular insect "species" beyond everyday usage. The compilation known as the Cyranides distinguishes different kinds of ants, which was never done by Aristotle or later natural historians despite a great interest in their habits (Beavis 1988: 200). There is also a recognizable classification of forms of the Egyptian sacred scarab. As well as being used in charms and amulets, the scarab had an important role in the "applied science" of theurgy, whose practitioners distinguished different varieties with particular properties (Beavis 1988: 159–60). They figure prominently in various charms and spells described in the magical papyri to invoke the aid of and gain power over the deities associated with the sun and moon. For each deity the appropriate variety of scarab, *heliakos* (sun scarab), or *seleniakos* (moon scarab) is prescribed. These two are individually described in the *Hieroglyphika* attributed to Horapollo, along with a third, the *ibiomorphos* (ibis scarab), with a single slender backward curving horn shaped liked the bill of the bird in question.

For all its undoubted successes, Aristotle's work also has severe limitations. He works almost all of the time with a pre-existing selection of species for discussion, essentially those that were of interest to practitioners or those few with a place in folklore and literature. There is no desire to create a comprehensive fauna of the Greek world, or any attempt to document all the creatures which an objective survey of the Greek countryside might produce. One might have expected the "father of zoology," if such he was, to "discover" and describe for the first time many new species, but this is not the case. There seem to be only about five invertebrates described by Aristotle for the first time: these include the pseudoscorpion or book-scorpion, the *xylophoron* (bagworm moth larva), and the *ichneumon* (spider-hunting) wasp (Beavis 1988: 34, 148–9, 189). Although Aristotle declares an intention of not ignoring even the most insignificant creatures, there are in fact several invertebrate types well known in popular lore which he either only mentions in passing—such as leeches or the *ioulos* (millipede or woodlouse)—or ignores entirely—for example, the cockroach, praying mantis and rose chafer beetle. Aristotle makes no attempt to deal with those curiously "missing" invertebrate types such as dragonflies or the many attractive butterflies besides the one pest species he describes. In fact we have to wait until the seventeenth century for that step to be taken: the *Theatre of Insects*, though largely based on Aristotle's range of invertebrate types, does include dragonflies and begins the process of distinguishing different butterfly species by their wing markings (Mouffet 1658: 939–43, 957–75).

Aristotle's handling of the subject of spontaneous generation—the belief that invertebrates are produced by the natural transmutation of various substances—is curiously inconsistent. Although he rejects this concept in cases where he knows for certain that a particular insect has a continuous life cycle and reproduces sexually, he falls back upon the traditional theory whenever contrary evidence is unavailable (Davies and Kathirithamby 1986: 21–2). His acceptance of the principle of spontaneous generation is not challenged by the few instances where he has become convinced that it does not occur, and therefore it continues to present a major obstacle to the study of most insect life histories, curtailing curiosity by the assumption that there is nothing further to discover. Spontaneous generation thus serves to disguise, even from Aristotle himself, vast areas of ignorance.

In popular belief all insects were spontaneously generated, and even in cases where Aristotle could demonstrate otherwise, the older view refused to be eradicated. For every type to which Aristotle attributes sexual reproduction and a continuous life cycle, an alternative folk belief involving spontaneous generation can be found. Aristotle knows that scorpions reproduce sexually, but others claimed they were generated from dead crabs (Beavis 1988: 26–7). Similarly, Aristotle can describe the life history of grasshoppers or locusts, and of cicadas, while folk tradition believed these to arise from animal and human corpses and from the earth (or from cuckoo spit) respectively (Beavis 1988: 69–71, 96–7). But in Aristotle and Theophrastus these instances of a "scientific" viewpoint are outnumbered by cases where the older belief is still uncritically accepted, even in cases where Aristotle can provide quite a detailed description of a life history. For example, even though he can detail the development of cabbage white butterflies from egg—although he doesn't recognize it as an egg—to adult, he still believed that the former is spontaneously generated on the foodplant (Beavis 1988: 125–6). Aristotle has a curious category of spontaneously generated insects which he believes do mate but then give rise not to creatures like themselves but only to "something imperfect," from which nothing further is produced. Thus, he fails to recognize nits as the eggs of lice, giving rise to miniature versions of the adult, and he attributes a similar incomplete life cycle to the domestic fly (Beavis 1988: 114, 220).

The Aristotelian tradition shows little evidence of the systematic study of invertebrates in the field. Aristotle and Theophrastus are engaged in a primarily literary and philosophical exercise, building theoretical structures into which concrete facts are drawn by way of support and illustration. Where data from practitioners in the various technical fields is unavailable, they present the appearance of passive rather than active observers, with a tendency to draw hasty conclusions from insufficient information, and to fill gaps in their knowledge by speculation rather than planned empirical research. Thus, in the *Generation of Animals* (759a–760b) Aristotle has a long discussion on various theories of the origin of the three castes of honey bees which he has been presented with by his beekeeper informants. Although he expresses the principle that observation must be the final arbiter of truth, he in fact makes his decision on purely theoretical grounds. He first reviews the possibilities that they fetch their offspring from elsewhere, these being either spontaneously generated in flowers or the offspring of some other creature; or that they generate the young themselves; or that they fetch some and generate others. Aristotle concludes that the known facts about bees, combined with general considerations drawn from the reproduction of animals in general, make all these theories impossible. If, he argues, they get their offspring from elsewhere, we ought to see bees spontaneously generated in the wild, since there is no reason why bees should develop if the seed is gathered up but not if it is left in situ. Furthermore, it makes no sense for an animal to collect another's offspring: animals only care for what at least appears to be their own. This is sensible enough, but Aristotle then misses the truth by ruling out sexual reproduction, partly on the basis of misinformation from beekeepers and partly because of over-confident reliance on theoretical principles. He argues, for example, that the workers (as we would call them) cannot be female, as nature does not give defensive weapons to any female creature; equally it is unreasonable to say that the workers are male, as no male creatures take trouble over their young. Aristotle's conclusion is that the only possible option is that bees are sexless, and that the workers generate the drones without mating, while the kings generate their own kind and the workers in the same fashion.

Although Aristotle states that in principle observation should take precedence over theory (*Generation of Animals* 760b), it is often the case that his theoretical beliefs override and determine his interpretation of his actual observations, which are explained so as to fit in with the former rather than being allowed to challenge them. He tries to harmonize the metamorphosis of insects such as butterflies with the development of higher animals by making the pupa correspond with the egg, and making the earlier stages the egg in process of development (Beavis 1988: 125). Even in the case of eggs produced by sexual reproduction, Aristotle refuses to term them as such, insisting that they are really *skolekia* ("little grubs"). Interestingly enough, Pliny, who has no overriding theory constraining him, does talk about insect eggs and is thus more "correct" by modern standards. Thus, we have the contents of the spider's egg mass, quite a conspicuous object even to the casual observer, described as being tiny larvae that are round at first and subsequently take on an articulated form; which strongly suggests that Aristotle has spotted an egg mass one day and some while later found in its place a cluster of tiny spiders, bunched together as the original eggs had been (Beavis 1988: 38).

Although Aristotle and Theophrastus do sometimes challenge, correct, and improve upon folk traditions and the lore of farmers, doctors, and other practitioners, there is no consistent effort to seek out and confirm the truth in all cases. As with the question of spontaneous generation, much erroneous information is incorporated without question. The popular identification of queen bees as "kings," based on human social roles, is accepted, and the identification of worker bees as neither male nor female is based on generalizations about the nature of each gender (*Generation of Animals* 759a). Aristotle's apparent success in declaring that only male cicadas sing is taken over from folk tradition, which happened by chance to be correct: Aristotle had no method of sexing insects. Theophrastus accepts without question farmers' conviction that the action of the fig-wasp is necessary for the production of the fruit, and he endeavors to explain away apparent exceptions by theoretical speculation as to the effect the wasp was supposed to have on the developing fruit (Beavis 1988: 212–14).

A number of factors probably held Aristotle back from establishing a truly "scientific" study of invertebrates, or of animals in general. His predominantly literary and cerebral culture lent a definite superficiality to his study of the facts of the real world. Although he claims to depart from Plato in beginning with the concrete particulars of the world, he is not really so far distant, and he is generally rather too hasty in leaving behind the actual lives of insects in favor of theoretical constructs. Furthermore, Aristotle lacked the single-minded dedication that would have been required to create a truly scientific entomology. With his diverse interests, he was impeded from being more than a compiler and casual observer of the natural world. It is instructive to compare Aristotle with the Aurelians, the early entomologists of Britain, who with no more resources available to them than he had collected and cataloged species in a systematic fashion and bred them in captivity to determine their life cycles.

Perhaps too the gulf between cultured society and the trades and professions inhibited Aristotle from this kind of practical research.

Although limited in many respects, the Aristotelian achievement is nevertheless in outstanding contrast to the work of later authors, who admire it to the point of reverence but make no attempt to build upon it. It is notable that there is no progress on the study of life histories: the only new ones to appear belong to the Chinese silk moth—which was not known to Aristotle, his silk moth being a native Greek species—and the Egyptian scarab, where the primary interest is religious rather than zoological (Beavis 1988: 143–5, 160–1). For Aristotle, the scarab is a simple member of the Mediterranean insect fauna, but subsequent texts begin to be influenced by Egyptian lore, in which it was a sacred creature, associated with the sun, the initial creation of the universe, and its recreation each new day. It is the poor quality of what follows that makes Aristotle seem "scientific" by comparison. There is some evidence that later followers of Aristotle other than Theophrastus did continue his researches in somewhat of the same spirit, since some of the non-Aristotelian material in the *History of Animals* is of quite high quality: there is, for example, a fresh treatment of the subdivisions of spiders, bees, and wasps, and a good description of the construction of the orb web (Beavis 1988: 36–7, 40–1, 188–9, 195–8). But this tradition was evidently short-lived, and entomology and zoology failed to achieve a separate existence as an intellectual discipline. This failure was due to the fact that zoology had no obvious "function" in either practical existence or educated culture, and no obvious professional "base." The exchange of information between the technical traditions and the high culture of philosophy had been all one way. There was no conception of the value of an "applied entomology," whereby insights from a "scientific" and objective study of insects could be applied to the solution of practical problems, so enabling farmers to better control the pests that infested their crops, instead of relying on a combination of magic, prayer and "remedies" of dubious efficacy.

What we do find in succession to Aristotle is a tradition of "natural history" writing, which is far from scientific but arises partly from an impulse to entertain and instruct the literate public, and partly from an encyclopedic impulse where the accumulation and cataloging of literary subject matter becomes an end in itself, more or less divorced from contact with the actual subjects in the real world. Aristotle and Theophrastus are revered as authorities, and the data they provide are constantly reproduced in new forms, but they are so revered as to help stifle new discovery, since whatever they declare to be the case is accepted as the last word on the subject, not available to be challenged or supplemented. This is particularly ironic since Aristotle's work always acknowledges the principle of the primacy of observation over theory and hearsay, and since a work like the *History of Animals* presents a rather unfinished appearance, is not entirely internally self-consistent, and is in part a collection of notes rather than a polished treatise.

The growing antiquarianism of Greco-Roman culture, and its increasingly total reliance upon written sources, meant that the written word was invested with excessive authority and credited with virtual infallibility. Where literary sources conflict, a Greco-Roman scholar will prefer to set both side by side rather than attempt to decide, let alone discover for himself, which is correct. Furthermore, most ancient writers on natural history make no distinction between the authority of a serious student of animal life such as Aristotle, and that of a poet or dramatist who may mention insects by way of metaphor or background detail. This can be seen, for example, in the lists of authorities provided by Pliny in book 1 of his *Natural History*. There is no question for Pliny of taking Aristotle as decisive over a debated issue such as whether the king bee has a sting. If anything, literary sources tend to be given more authority because of the greater scholarly attention devoted to them. Homer, in particular, is widely quoted by natural historians, as a direct consequence of the reverence in which the Iliad and Odyssey were held on account of their antiquity. These were perhaps the nearest thing that Greco-Roman paganism had to sacred scripture. Pliny and Aelian often quote Homer to illustrate aspects of natural history. Pliny regards Homer as the greatest intellect of all time, outstanding not only for his literary ability but for his subject matter, and he describes his writings as the "the most precious work of the human mind" (*Natural History* VII.107–8). Pliny is also a great admirer of Virgil and frequently cites his didactic poem on agriculture the *Georgics* (*Natural History* XIV.7, XXIX.28). Although Virgil did not intend his poem to be a serious work of instruction, the poet's fame makes it an authority on the natural history of the bee and on practical apiculture. In a sense all these authors coexist in a timeless continuum, utterly unlike our modern concept of more or less up-to-date authorities.

We might also say that lack of progress in knowledge is not unexpected when there was uncertainty over the very concept of progress. The ancients did not only have difficulty in distinguishing between reliable and unreliable authorities, or between serious "science" and literary fancy. Alongside the general lack of awareness of authors' historical context—that "timeless continuum"—there exists, if anything, a belief in regression in knowledge, produced by reverence for the writings of antiquity, the older the better, with Homer and Hesiod at the head, along with mythical authors such as Orpheus who, according to Pliny, was the first to write on herbal medicine (*Natural History* XXV.11–13). All knowledge, it was widely believed, had been possessed by the ancients, and if anything, there was degeneration in times since. Although it was understood that the human race had arisen from primitive conditions to civilization, civilization was understood as a more or less static achievement, existing already in the time of Homer (*Natural History* XII.1–2). Pliny states that a thousand years ago at the dawn of literature Hesiod gave instructions for agriculture; these, he implies, have not been improved upon. Indeed, he continues, much of what the ancients knew has been lost through our neglect and has had to be rediscovered—a principle which would, of course, enable innovations to be reinterpreted as rediscoveries (*Natural History* XIV.3–4).

Greek achievements up to and including Aristotle had led by Pliny's time to a loss of confidence. Reverence for the ancients led to feelings of inadequacy and loss of a desire to emulate. Pliny, having read all the available literature on the subjects with which he deals, has noticed that later authors tend to copy older ones word for word (*Natural History* Preface 21–2): he, of course, does the same, but is careful to name his authorities. Pliny tends to exaggerate the achievement of the ancients, and to imagine they have surveyed the ground so comprehensively as to leave little room for their successors. He gives, for example, a fanciful account of how Alexander organized a supply of data for Aristotle's researches by giving orders "to some thousands of persons throughout the whole of Asia and Greece, all those who made their living by hunting fowling and fishing, and those who were in charge of warrens, herds, apiaries, fishponds and aviaries, to obey his instructions, so that he might not fail to be informed about any creature born anywhere" (*Natural History* VIII.44). Nowadays, Pliny complains, there is no fresh research or effort to help human life (*Natural History* XXV.2). However, he does not attempt to fill that need himself, nor does he blame that adulation of antiquity which he himself exhibits.

Where Aristotle is used, he is often misinterpreted. Aristophanes of Byzantium's epitome of the *History of Animals* introduces confusion by for example conflating originally distinct life history accounts—such as those for butterflies and silk-moths—into one (Beavis 1988: 124, 126, 143). Pliny's *Natural History* contains numerous examples of ludicrous misunderstandings of Aristotle's text, which some empirical observation would have enabled him to clarify, along with abbreviated summaries that obscure the original sense, and conflation of separate accounts into one composite description (Beavis 1988: 69). There is also the phenomenon of the apparent interpolation of Aristotle's works with spurious material, odd scraps of folklore, and fabulous accounts like those of the "snow worm"—a creature supposed to be spontaneously generated from long fallen snow—and the "fire born" which was reputed to be generated from fire and to flourish only within this element (Beavis 1988: 247–8). Interpolation has also introduced confusion into some of his life history descriptions, and added statements that contradict the original text, such as the assertion that honey comes from the air rather than being made (as the original text states) by bees from flowers (*History of Animals* 553b–554a).

In keeping with the impulse to entertain, a good deal of attention is devoted to "wonders" which in the case of entomology relate to fanciful tales of the behavior of known insects or extravagant accounts of exotic species from the fringes of the known world. A literary tradition, paradoxography, develops, including one work attributed to Aristotle, which simply catalog marvels without any attempt at discrimination or verification, and similar material is found in natural historians such as Pliny, the geographers, and in the accounts of Alexander's eastern conquests, which provided a major impulse for the development of such tales. Wonder tales are copied from one author to another, embellished and retold in different forms, as the many divergent accounts of the fierce Indian ant that excavates gold in the process of making its nest burrows bear witness (Beavis 1988: 209–11).

The post-Aristotelian natural history tradition continues to propagate old errors from folk tradition which Aristotle had disproved. The idea of silk as a plant product continues to occur, despite Aristotle having established that it is produced by a caterpillar (Beavis 1988: 145). This happens especially in the case of those species which Aristotle discovered reproduce sexually, but where the older concept of spontaneous generation carries on nonetheless. In the case of the cicada, authors such as Plutarch and Lucretius still assert that they are born from the earth (Beavis 1988: 96–7). In the case of Lucretius, who has some claim to a scientific approach to natural phenomena, this failure is perhaps surprising, but Lucretius, in some ways like Aristotle, is more concerned with the use of facts to buttress his philosophical scheme than with verifying the accuracy of those facts: any explanation will do, provided it meets the needs of his argument, in this case that no supernatural agency can be discerned in the natural world. By the time of Isidore on the threshold of the Middle Ages, Aristotle's discoveries have entirely vanished from view, and all invertebrates are once again regarded as spontaneously generated.

The post-Aristotelian natural history tradition also reverts to popular anthropomorphism, and in attributing human emotions and mental processes to animals employs them to provide moral lessons for human life. The most popular subjects among insects are the ant and the honey bee, whose life in community was held up as a model of peaceful coexistence, subordination of self-interest to the common good, and hard work and dedication. Ants in fact form a conspicuous part of a philosophical debate about animal intelligence, but this is really a debate about man and his superiority or otherwise to animals (Beavis 1988: 206–7). Although there is evidence of authors watching ants, there is no systematic study of their habits; for example, no one had seriously investigated the internal layout of ants' nests, but various assumptions were made about them and a picture built up largely from guesswork, and much folk belief was uncritically accepted (Beavis 1988: 201–2). Post-Aristotelian literature on honey bees contrasts with his work by postulating a system of government, political intrigues, warfare, and other analogues of human behavior (Davies and Kathirithamby 1986: 59–63). Pliny and Virgil describe the bee colony as operating like a military encampment, with, for example, one bee whose task it is to wake the others in the morning with a special call, on the instructions of the king, and to fly around making a similar call when it is time to stop work and go to sleep. There is said to be a division of labor organized by the king, who goes around encouraging the workers. Bees that are idle are marked out and punished, even with death for the incorrigible. One persistent theme in Aelian's natural history is that the behavior of creatures inferior to man and yet often displaying superior virtues can provide man with a salutary moral challenge. Aelian has a message about the untaught virtues of irrational animals being a lesson for decadent mankind. Historians, he says (*On Animals* I.59), celebrate the great palaces of mighty potentates like Cyrus and Darius, but ignore the more skilful buildings created by bees. Moreover, these great emperors accomplished their works through the suffering of many, but the bee is the most kindly and wisest of creatures.

For authors such as Pliny and Aelian, it is not accidental that animals are moral exemplars, for this is one of the purposes for which they exist, just as it is not accidental but providential that the bee is so useful to mankind. Honey for Pliny is a major gift of beneficent Nature, who made the bee to provide it for us (*Natural History* XXII.107). Numerous animals and plants are described by Pliny as providing evidence for his belief that Nature works for the good of mankind, although humanity is itself a part of nature. Thus, glow worms are characterized as terrestrial stars, created to provide weather signs to farmers (*Natural History* XVIII.250–3). Even bed bugs are medicinally useful, for "the universal mother gave birth to nothing without very good reasons" (*Natural History* XXIX.64).

These moral messages relating natural history to human lifestyle are not casual and irrelevant asides but are one of major purposes underlying such treatises. Knowledge of the natural world is rarely acquired and propagated simply for its own sake, following the truth wherever it might lead out of pure curiosity: the practical implications for human life are rarely far from view. Farmers and technical practitioners might be able to restrict themselves to practical skills—as Columella (*De Re Rustica* IX.2) says—but literate culture could not separate life and thought into discrete compartments. What we call science and morality are naturally interlinked, because morality is seen as not being a merely human construct, but as part of the fabric of the universe. In fact, there are no purely human constructs since humanity is part of nature and nature governs human life. Everything man does is based on the resources of nature, which is the logic behind the appearance of human creativity in Pliny's *Natural History*.

Literary interests also play a significant role in the tradition of natural history writing that follows Aristotle. The evidence of surviving fragments in commentaries on literary texts, dictionaries, and later compilers such as

Pliny strongly suggests that the lost Hellenistic "zoological" writers such as Sostratus and Pamphilus did not made any advance on Aristotle in terms of real entomological knowledge and research. Such Alexandrian scholarship was overwhelmingly literary, devoted to editing, commenting upon, and compiling data from pre-existing texts. So far as the study of insects is concerned, its prime motive was probably lexicographical. Insect names were collected for dictionaries and encyclopedias such as that of Hesychius, and to explain allusions in the works of poets and dramatists. Although lexicographers could have made some progress in nomenclature, they normally confuse rather than enlighten, with vague definitions ("a certain animal" is one that frequently occurs) and utterly contradictory definitions set side by side. This lexicographical and expository tradition is demonstrably out of touch with the concrete reality of insects and their behavior. While, for example, there is some interesting discussion about the praying mantis, it is solely aimed at elucidating a certain passage from the poet Theocritus (Beavis 1988: 85–8). Similarly, the gold-green rose chafer beetle, played with by children, features in literary works by Aristophanes and others and gives rise to an exegetical tradition exploring its identity and relationships with similar creatures (Beavis 1988: 164–8). Literary commentators tend to rely on guesswork and deduction from the text, rather than on objective investigation of facts. Hesychius plainly has no real idea of the meaning of most of the obscure insect names he has assembled, many of which are in fact dialect synonyms of well-known types such as the cicada. That the work of supposed Hellenistic zoologists was along these lines can be illustrated by the fragment of Sostratus where he attempts to distinguish two terms for horseflies and other cattle-biting insects by meaningless criteria, entirely unaware that in the literary sources in which he finds these names they are synonymous (Beavis 1988: 226).

Pliny's vast compilation, drawing material from all available sources including Aristotle and Theophrastus, demonstrates clearly the failure to progress. No development has been made in the study of life cycles, and there is no evidence of the discovery and description of new species beyond the basic fauna established by the technical traditions, Aristotle and Theophrastus, and the geographers. Pliny himself contributes very little, although he does add a handful of insect names from Roman folk tradition, such as the edible wood-boring larva known as *cossus*, and the stag beetle used as an amulet (Beavis 1988: 152, 154). He also adds a few fresh observations about pre-existing types such as early spring butterflies and the glow worm. Working mostly from pre-existing sources, Pliny often misinterprets the material he is reproducing, he multiplies species unnecessarily by failing to recognize synonyms, and confuses creatures with similar names—as when he muddles the silkworm *bombyx* with the solitary bee that Aristotle calls *bombykion* (Beavis 1988: 197). By his uncritical compilation of material of varying quality and often self-contradictory nature, Pliny establishes a model of indiscriminate and undiscriminating compilation which is followed by Isidore in late antiquity and into the Middle Ages and survives as late as Mouffet's *Theatre of Insects* in the 1600s.

After Aristotle, in the absence of an applied entomology, the technical traditions go their separate ways, becoming increasingly corrupt and derivative. The Latin agricultural writers do, it is true, represent a fresh engagement between literary and mundane culture, but they have no fresh contribution to make to the knowledge of the insect pests with which they deal, or to the natural history of the domestic honey bee: Columella's long account shows no real interest in the question of how bees reproduce (*De Re Rustica* IX.2). The trend is generally towards massive compilations of second-hand material such as we find in late antiquity with the Geoponica on agriculture and the Hippiatrica on veterinary medicine, where information from diverse sources is set down in a repetitive fashion with no attempt to sift or analyse it, or to advance knowledge beyond what earlier authorities have declared. Pharmacology remains with the same limited range of insect types, mostly those readily available in domestic contexts.

The rise of Christianity, too, had its effect upon the study of insects outside the technical traditions. Christianity followed the Stoic line on the purpose of animals, as is illustrated by the debate in Origen's *Contra Celsum* where the theologian combats Celsus' view that the world is designed for animals as much as for man, and that animals display a rationality not inferior to that of human beings (Beavis 1988: 206–7). The specific animal used as the basis for discussion is the ant, with its reputation for human-style communal life. The Christian worldview made theology the center of intellectual activity, as philosophy had been before, and still provided no place for the study of animals as a matter of interest and curiosity in their own right. The existing anthropocentric bias is confirmed, and the range of insects dealt with in literature contracts rather than expands.

Since the majority of Christian literary activity was devoted to the exposition of Scripture, it is not surprising that attention is focused mainly on a narrow group of insects referred to in Scripture. In Christian times, the Bible acquired a similar role to Homer or Virgil in the study of natural history, with a disproportionate amount of effort being devoted to the study of animals as described in Scriptural texts. These are partly creatures such as the ant from which moral lessons could readily be taken, and partly those more obscure ones—such as the various types of locust rendered in curious ways by the translators of the Greek Old Testament (Beavis 1988: 63–9)—whose nature needed to be explained in order to make sense of certain difficult passages. As with the lexicographers of earlier times, this tradition of biblical commentary operates entirely within the bounds of the text and does not touch concrete reality at any point. The curious choices of Greek words by the Septuagint translators to render Hebrew animal names whose meaning they really had no idea of, were made the basis of fanciful speculations, most famously with regard to the *myrmekoleon* or "ant-lion"—not to be confused with the real-life ant predator of modern entomology—which was said to be a hybrid between ants and lions. Much of the material in late popular natural history compilations like the Physiologus, ancestor of the mediaeval bestiaries, derives from exegetical material such as the commentaries of the Fathers on passages of Scripture that happen to mention creatures such as locusts or beetles. This is how the scarab's symbolical life history comes to be described by prominent theologians such as Clement of Alexandria, Eusebius, and Augustine of Hippo (Beavis 1988: 161). Isidore's encyclopedic *Origines* reflects more the tradition of Pliny but presents the surviving remnants of the ancient knowledge of insects in a sadly degenerate form. Also sadly degenerate are the glossaries of late antiquity, which show how insect nomenclature has become much vaguer and more imprecise.

Beyond the Classical period, the minimal study of insects (and other animals) that still existed in the Latin West and the Byzantine empire remained firmly in the traditions of Pliny and Aelian respectively. The revitalization of the Aristotelian tradition remained in abeyance until the 1500s and 1600s when writers like Mouffet—if not breaking free entirely from Pliny's veneration of the wisdom of the ancients—did at least seek to build upon it and prepared the way for the birth of modern scientific entomology in the 1700s. Thus Aristotle belatedly became what he never aspired to be, the "father of entomology."

CHAPTER TWO

Insects and Disease

Three Insects That Affected the People of Ancient Egypt

JOYCE M. FILER

The aim of this chapter is to consider how three particular insects—bees, flies, and mosquitoes—affected ancient Egyptian society, with particular reference to their effect on the health of the populace.

In the modern world it is recognized that some insects are a serious factor in the spread of disease; "approximately one in six people alive today is currently affected by an insect-borne disease" (McGavin 2000: 6), but, obviously, it is difficult to apply such statistics to the people of ancient Egypt. The ancient Egyptians were keen observers of their environment and were aware of the insects around them (Curry 1979: 113), and so it may well be that they did connect certain insects with the dissemination of disease, however, like many ancient societies, it is likely they blamed any ill health on the will (and displeasure) of the gods (Conrad 1995: 91; Longrigg 1995: 30; David 2002: 286). The sources for this consideration of three insects which influenced the lives of ancient Egyptian and Nubians will include: modern knowledge of the insects and the diseases they carry, ancient textual and artistic references, and biological information from skeletons and mummies. Where relevant, examples will be drawn from various periods of ancient Egyptian history—from pre-dynastic times, the pharaonic and Graeco-Roman periods through to the Islamic/Medieval period—in order to illustrate the longevity of the influence of these three insects on ancient Egyptian and Nubian societies.

Researchers point out that, whilst the development of farming has its advantages in allowing population growth, an unfortunate consequence of changed human settlement patterns is the development of new infectious diseases (Cohen 1989: 39; Filer 2001: 133; Porter 2003: 4; Lieberman 2013: 197, 199); the development of animal husbandry also brought people into contact with various diseases. Insects, and the diseases they transmit, played their part in these human activities.

Two of the insects in this discussion, flies and mosquitoes, certainly deserve to be called, respectively, "the bad" and "the deadly," the "success" of both of these insects being "encouraged" by human alteration of the landscape. Settlement living resulted in the accumulation of human refuse, especially feces, thereby providing ideal conditions for flies and the diseases they spread. "Haunting agricultural settlements" (Porter 2003: 6), mosquitoes were greatly aided by human land clearance and so the incidence of malaria, which has probably caused more human deaths than any other disease (Cohen 1989: 41), increased. This chapter, however, will start on a more positive note with an insect that greatly aided the health and wellbeing of the ancient Egyptians.

BEES: BENEFICIAL TO HEALTH

As will be seen in this section bees made a valuable contribution to ancient Egyptian society. Bees were so important to the ancient Egyptians that the insect (together with the sedge plant) formed a part of one of the King's titles, the prenomen (Figure 2.1). Here a bee motif (together with a sedge-plant motif) is placed before the prenomen title, which is enclosed within a cartouche, signifying that the King is "he of the sedge and the bee" and by enclosing his name it may well be that the King is indicating that "he was the ruler of all 'that is encircled by the sun'" (Quirke and Spencer 1992: 227).

In considering the contribution these remarkable insects made to the life of the ancient Egyptian populace, and to health issues in particular, it might be useful, first of all, to consider their value today. In their series entitled "Unsung heroes" the Royal Society for the Protection of Animals (RSPCA) (2019: 3) describe the honey bee as one of "the less 'cute and cuddly' animals all around us" but goes on to stress how important these "amazing creatures" are to our world. How did the ancient Egyptians regard these "unsung heroes"? The available evidence indicates that bees were greatly valued in ancient Egyptian society and bee products, as we shall see, certainly featured significantly in the world of Egyptian medicine and health. Following Kritsky (2015: xiii), the term "honey bee," as opposed to "honeybee" will be used here. Honey bees themselves offered no discernible disease threats to the people of ancient Egypt, and, apart from the inconvenience of being stung, the ancient Egyptians benefitted greatly from their association with these insects.

FIGURE 2.1 A painted lintel across columns in the Hypostyle Hall, Temple of Karnak, Nineteenth Dynasty. A bee is featured, at the left, before the King's cartouche.

Honey bees produce a number of products—honey, wax, royal jelly, propolis, and venom—all of which benefit human beings; however, this discussion will concentrate on honey and its contribution to Egyptian medicine. By beating their wings bees are able to create airflows in their hives thereby lowering the water content of their honey which, in turn, thickens the product. This very low water content together with its low pH measure inhibits the growth of bacteria and so honey was a valuable commodity in the fight against infection.

As we shall see, honey was a very popular ingredient in medical recipes, appearing at least five hundred times in various ancient Egyptian texts (Zumla and Lulat 1989: 384); honey is also mentioned, as a medicant, in everyday ancient Egyptian correspondence. A poignant letter from ancient Egypt is that from the draftsman Pay to his son Pre[emhab], also a draftsman. Written during the reign of the pharaoh Ramesses II, the ailing Pay is begging for his son's help: "Don't turn your back on me, for I'm not well." Pay reveals that he is blind: "Don't be sparing in tears for me because I am in this darkness (i.e., blindness)"; Pay believes the god Amon has deserted him. Pay asks his son for medicine: "May you bring me a bit of honey for my eyes as well as ochre ... and genuine galena" (Wente 1990: 142). We don't know the cause of Pay's blindness but it is clear he has great faith in the soothing powers of honey for his eye complaint. In a humble way it is a testament to the healing power of the honey bee and its honey. More eye complaints are described below in the section on the housefly, and it is likely that honey was used to relieve the often-painful symptoms inflicted by these insects.

There is a large corpus of medical texts from ancient Egypt containing "recipes" aimed at curing various ailments and health issues. Of particular relevance to this overview of honey in medicine is Papyrus Ebers (*c.* 1534 BCE)—writings on general medicine, gastrology, urology, and gynecology—in which honey is frequently mentioned as an ingredient. Readers might like to refer to the translations by Bryan (1930) and Ebbell (1937), although caution is advised when using the latter (Nunn 1996: 30). Also relevant is the Edwin Smith Papyrus (*c.* 1550 BCE), translated by Breasted (1930), which deals with surgery and in which honey plays a part. Other texts that mention honey include Papyrus Berlin 3038 (*c.* 1200 BCE) on general medicine, Papyrus Chester Beatty VI (*c.* 1200 BCE) on rectal diseases, and the London Medical Papyrus (*c.* 1300 BCE) concerning mainly magical issues.

A survey of the medical texts indicate that honey was thought to promote healing—either through being used as a binding agent or as a prominent ingredient—in a variety of ailments and life-changing events in Egypt. These texts include recipes recommending a variety of substances to be mixed together and if, as in many cases, the mixed ingredients were to be formed into a pessary then honey was often included as the binding agent and in many cases a paste was formed for application. With relation to gynecological issues, honey, as an ingredient in a pessary, is mentioned in recipes to both prevent conception and to induce an abortion. The viscosity of honey may have impeded the movement of sperm when used as a contraceptive (Robins 1993: 80). Monserrat (1996: 200), discussing Graeco-Roman Egypt, states "honey is a frequent ingredient in Demotic texts when a cloth pessary, soaked in honey (or honey and/or resins)" was required; such a pessary was inserted into the vagina as a mild spermicide (Antelme and Rossini 2001: 133). To bring about an abortion, P. Ebers suggests forming a pessary comprising dates, onions, and acanthus fruit bound with honey—the concoction was to be applied to the vulva on a cloth (cited by McLaren1990: 29). However, in order to induce childbirth, P. Ebers 801 recommends that a strained mixture of honey and fenugreek should be drunk for one day. Fenugreek has been traditionally used to induce lactation (Simmonds et al. 2016: 198) and so its inclusion in a childcare-related situation is not surprising.

As well as being an aid to gynecological health concerns, honey appears as a component in medications to relieve a wide range of ailments concerning other parts of the body; the following examples will illustrate how valued honey was in the ancient Egyptian "healing arts."

Concerning problems with dental problems, Ghalioungui (1963: 57) suggested that lesions in the teeth were filled with a mixture of honey and

Nubian earth and cites P. Ebers 739, however, Miller (2008: 70) points out that this practice has not been confirmed in ancient Egyptian dentitions nor in other interpretations of the Ebers Papyrus; further, Nunn (1996: 205) states that honey would not have been helpful in cases of carious lesions. It is also important to note that before the introduction of sugarcane into Egypt, tooth caries is not frequently seen in dentitions of the pharaonic period. P. Ebers 741 suggests a mixture of figs, beans, malachite, ochre ground to a powder and fixed with honey for toothache and, in the event of blood being spat from the mouth—perhaps as a result of tooth or gum infections or even trauma—Papyrus. Chester Beatty 233 suggests a concoction of honey with acacia leaves, chicory, wild fennel, and other ingredients. Concerning maladies of the throat, Simmonds et al. (2016: 24) state that the root of the marshmallow plant was used in the preparation of throat lozenges in ancient Egypt, so it is likely that honey was used as a binding agent here. The ancient Egyptians chewed pellets of honey and aromatic spices to freshen their breath (Filer 2001: 36).

In dealing with chest and lung problems and coughs, a number of remedies involve honey and certainly honey would have had a soothing effect. P. Berlin 31 recommends honey with cumin and milk; a small cup inscribed with the (translated) words "cumin, set milk, honey" is described by Kritsky (2015: 93). Interestingly, cumin, thought to be native to Egypt (Simmonds et al. 2016: 67), is being considered nowadays to treat lung conditions, especially asthma. P. Berlin 40 suggests honey with acacia leaves and beer whilst Berlin 117 adds ochre to acacia leaves and honey as medicants. To treat catarrh, P. Ebers 63 recommends anointing with a mixture of honey, stibium, aloe, and dry myrrh. Ebers 318 and 234 both suggest the application of emmer grain flour or a wheat porridge with honey, together with goosefat in the former and ochre and figs in the latter. Figs, with honey, ochre, and gum, also feature in P. Ebers 57 as a cure for lung problems. Kamal (1975: 79) notes that P. Berlin 18 suggests a dressing combining honey and colocynth used to treat an abscess in the breast.

Herodotus Book II (1988: 158) tells us that the Egyptians "every month for three successive days they purge themselves, for their health's sake, with emetics and clysters, in the belief that all diseases come from the food a man eats ..." and certainly the medical papyri are a testament to the Egyptian concern with disorders of the stomach, and the urinary and defecatory systems, and certainly honey features in medicines for disorders of these bodily systems. Honey, together with milk and barley water, "were used as stomach demulcants" (Campbell 2008: 227). Honey has been used in the modern treatment of infantile gastric problems (Haffejee and Moosa 1985; Zumla and Lulat 1989: 384) and it is likely to have been used for the same purpose in ancient times.

To alleviate "worms in the stomach" P. Chester Beatty 100 recommends boiling honey with purslane and cow's milk which should be drunk every day for three days; Simmonds et al. (2016: 152) describe purslane as a "global panacea" useful in reducing fevers and ridding the body of toxins which are possibly the "worms" of ancient Egyptian medical thought. Originally from Asia, the appearance of purslane in Egyptian texts might lead us to reconsider ancient Egyptian trading practices. Other texts deal with "worms": honey with wine and frankincense (P. Berlin 7) whilst P. Ebers 12 recommends the use of a strained liquid laxative incorporating honey, wine and "tiger nuts"—to be drunk for one day. A stomach remedy with honey, moringa oil, wine, and frankincense is suggested in P. Ebers 214. The remains of various actual worms have been detected in the examination of some mummies, as seen in the section on houseflies. Nunn (1996: 71–3) gives useful comments on the issues in translating and identifying types of worms in ancient Egyptian texts.

The ancient Egyptians have sometimes been "accused" of being overly concerned with the anus and associated problems—Nunn (1996: 90) states "the anus received detailed attention in ancient Egyptian medicine" and he cites P. Ebers 138–64 and P. Chester Beatty VI in particular; the title "Shepherd of the Anus" may indicate an important medical role in the Egyptian royal household. For anal pains, P. Berlin 163 and 164 suggest an enema comprising honey, moringa oil, and sweet beer. Colocynth, or bitter apple, is native to North Africa and has known purgative properties although caution is needed in its use as it can cause violent pain, even death (Simmonds et al. 2016: 59) and it is no surprise that it appears as an ingredient in medicines for the anal tract; P. Ebers 160 suggests powdered colocynth with marshmallow and, of course, honey "to cool the anus." Constipation seems to have been a concern as a number of recipes seek to alleviate the problem: the London Medical Papyrus 154 suggests an enema of honey, bean meal, oil or fat, and sweet beer applied to the anus for four days; P. Ebers 137 recommends juniper berries (along with frankincense, "tiger nuts," raisins, figs, sweet beer, persea fruit) and honey to ease constipation. Juniper oil may have diuretic effects and is used in remedies for kidney and bladder problems, flatulence, and digestive issues; however, again, caution is advised as these berries may irritate the kidneys adversely (Simmonds et al. 2016: 110). P. Ebers 89 also suggests a mixture of honey, juniper berries, and sweet beer to promote defecation.

Nowadays, honey has proved successful in the treatment of infected burns, ulcers, and surgical wounds (Zumla and Lulat 1989: 384) but the ancient texts indicate the Egyptians were already utilizing honey's curative powers for burns. P. Ebers 506 suggests using powdered carob pods mixed with honey and applied in the form of a bandage; P. Ebers 491 recommends treating an infected burn (i.e., one that is putrefying) with a dressing of honey combined with cinnamon, fragrant myrrh, turpentine, and sagapen. Painful mouth ulcers, termed "eating ulcers," were treated with a paste made from honey, cinnamon, gum, and oil or fat (Manniche 1989: 89). The third-century Demotic Magical Papyrus of

London and Leiden, Column XIX 32–40, recommends applying honey and rue to a wound resulting from a dog bite in conjunction with a particular spell being spoken over the wound (Griffith and Thompson 1904: 127). Bites were clearly an enduring problem as the Medieval author Al-Biruni (973?–1048 CE) recommends honey, together with various other ingredients, to cure the bites from crocodiles, monkeys, and humans (Kamal 1975: 95).

It is by examining the Edwin Smith (Surgical) Papyrus that we can best witness the value placed by ancient Egyptian physicians on honey as a healer of wounds. This papyrus was bought in 1862 by Edwin Smith, an American resident in Luxor, and hence named after him. It was once thought to be a New Kingdom copy of a much earlier document possibly dating to the Old Kingdom but this is now questioned (Nunn 1996: 27). The papyrus appears to be the notes of a doctor (or doctors) detailing how to deal with a variety of injuries, possibly those incurred during battles (Figure 2.2) but more "domestic"-type injuries cannot be excluded; the injuries are dealt with in a topographical

FIGURE 2.2 Battle scene illustrating how easily head injuries occurred, honey may have aided the healing process. Tomb of Anta, Deshasheh, Fifth Dynasty (after drawing after Petrie, 1898 by N. Parker). © Joyce M. Filer.

manner—starting at the top with injuries to the head and face, then the neck and so, assumedly, on down the body (however, the text is incomplete). Each injury is presented as a case wherein the injury is described, then a prognosis offered as to whether the patient will either recover, probably recover, or is highly likely to die; for the first two prognoses treatment is recommended. Various treatments are proposed including: surgical intervention and the application of a salve or fresh meat to the injury. Unfortunately, the Edwin Smith Papyrus breaks off halfway through case 48 describing a sprain in a vertebra. Majno (1975: 90–104) provides a useful discussion of some of the Edwin Smith cases from a modern medical point of view.

Of the forty-eight medical cases described in the Edwin Smith Surgical Papyrus six cases mention honey alone as the suggested treatment. Applied externally, the honey was likely to have been covered with bandages to prevent flies, attracted to the sweet-tasting substance, from settling on the wound (Breasted 1930: 101). In many cases a grease and honey ointment was recommended following surgery as part of the recommended medication. For example, case nos. 2, 10, 14, 23, 26, 27, 28, and 47 describe flesh wounds which are first dealt with surgically then followed by an application of a grease and honey ointment. Whereas for bone injuries, such as in case nos. 3, 11, 12, 25, 34, 35, 36 and 37, honey mixed with other ingredients was recommended following surgical intervention. Cases 3 to 9 describe very serious head injuries in which the skull bones are penetrated—the author/surgeon decides to treat some of them, the other cases are regarded as fatal and so no treatment is recommended. It is clear that honey played an important role in warding off infection in the treatable head injuries.

Scenes of war depicted on tomb and temple walls clearly show the range of injuries which might be inflicted during conflicts. For example, a Fifth Dynasty battle scene in the Tomb of Anta (variously written as Inti or Init) at Deshasheh shows close combat in which head injuries are being inflicted (Figure 2.2, drawn after Petrie 1898: pl. IV); these head injuries reflect some of the cases we encounter in the Edwin Smith Papyrus.

Another battle scene, from the Eleventh Dynasty Tomb of Khety at Beni Hasan (Newberry 1893: pl. XV) shows a siege scene involving archers; the piercing injuries and treatment, resulting from arrows, seem to be described in medical papyri.

The examination of archaeological skulls has revealed that some people can recover from even the most serious of injuries; in a survey of ancient Egyptian and Nubian head injuries examples of serious, yet healed, lesions were discovered (Filer 1992: 281–5; 1997: 63–5, 66–8). For example, in the Duckworth Laboratory, Cambridge, England, several ancient Egyptian adults presented bilateral, healed head injuries which had resulted in large “flaps” of bone being lifted from the skull. Again, the severity of such injuries is alluded to in the Edwin Smith Papyrus.

I would suggest that here the physician might have applied honey (possibly with other ingredients) to these injuries and then placed strips of linen (or papyrus) with beeswax over the wounds to hold the injured bone in place until healing had taken place. I believe honey probably made a major contribution to all the types in injuries seen on bones and depicted in battle scenes.

In their production of honey, it is evident that bees, our "unsung heroes," contributed greatly to ancient Egyptian life. Honey, as has been demonstrated, featured prominently in medications as an aid to improving health and defeating disease and infection.

We now turn to two insects, defined biologically as belonging to the order Diptera—flies and mosquitoes—which are the very opposite to honey bees in terms of their effects on human health in ancient Egypt and Nubia.

FLIES: FEARLESS AND FILTH-LOVING

Like mosquitoes (considered below), the common housefly *Musca domestica* is regarded with distaste. Like the mosquito, the housefly is also numbered as one of the most dangerous animals on earth—it may be small but this insect is capable of transmitting quite a formidable range of diseases to humans; although we are unlikely to see gross evidence of fly-borne diseases on ancient human remains, as is also the case with mosquito-borne infections (Hempel 2018: 122). However, through a perusal of ancient texts and artwork, together with our modern experience of houseflies, we can deduce the types of diseases flies transmitted to the ancient Egyptians.

Being keen observers of the fauna in their environment the Egyptians were aware of the behavior of flies, especially their persistence in returning time and time again to a desired source of food. On a carved inscription set up at Semna, King Sesostris III (*c.* 1860 BCE) deplores any lack of persistence in his soldiers when attacking the enemy:

> To stop when attacked is to make bold the foe's heart,
> Attack is valour, retreat is cowardice,
> A coward is he who is driven from his border
>
> (Lichtheim 1973: 119)

Thus, gold medals, in the form of flies, were given to persistent soldiers as a reward for their valor in battle and their prowess in pursuing the enemy (Curto 1971: 21; Kritsky and Cherry 2000: 61; Andrews 2003: 11; Morkot 2003: 86, 198; Shaw and Boatwright 2008: 37); a tomb inscription of Ahmose-pen-Nekhbet, a seasoned military campaigner under several Eighteenth Dynasty rulers, lists the rewards he received for bravery in the military field which included six gold flies (Morkot 2003: 4, 86). A necklace of large golden flies, found in the Eighteenth Dynasty, Theban tomb of Queen Ahhotep (Wilkinson 2005:15), may have belonged to her son Ahmose—a successful military king; however, the Queen may well have earned these "fly rewards of valor" in her own right as The Karnak Stele (set up by her son Ahmose) praises his mother's crucial involvement in military events (Robins 1993: 42–3; James 1973: 306).

Concerning biological evidence some remains of *Musca domestica* were found in the inner wrappings of Mummy 1770 (during the Manchester Mummy Project examination), however, as this mummy was re-wrapped (in antiquity) these are unlikely to be contemporaneous with the actual body but are certainly ancient (Curry 1979: 114–15, 116).

However, it is the behavior of flies with regard to their feeding and breeding habits that allows the transmission of disease. Even if the ancient Egyptians were unaware that flies caused infections, they were certainly aware of the symptoms of the diseases which we now know are potentially transmitted by flies: eye problems, dysentery, tuberculosis, typhoid, cholera, anthrax, and parasitic worms. As noted in the introduction, the house fly breeds in excrement and other types of waste. It transports bacterial organisms externally on its legs and after visiting fecal matter it then settles on human food and so bacteria are deposited when the fly cleans its head and proboscis. The fly also carries bacteria internally in its alimentary tract and this is deposited on human food when the fly, being unable to digest solid food, liquifies its meals and regurgitates it as "vomit." Another less-than-attractive behavior of the fly is its indiscriminate defecation on any surface, including human food items. Thus, the optimum conditions for the spread of disease is created.

On a personal note, years ago, whilst traveling on local transport through the western oases of Egypt, it was clear that flies were abundant even to the point of alighting on the eyeballs of an infant lying in its mother's arms next to me. That the child tolerated up to five flies at a time crawling in and around the eyes without complaining or reacting, and that the mother rarely waved the flies away, indicated the familiarity of these insects to the populace. This begs us to wonder how tolerated were flies in ancient Egyptian contexts? During those journeys into the oases (and other parts of modern Egypt) I saw much evidence of eye disease, likely caused by flies.

Eye diseases were also likely to have been very common in ancient Egypt and these potentially resulted in reduced sight, even blindness. The Ebers Papyrus includes references to at least twenty-nine eye complaints—some of which would have been caused by flies and at least ninety-two prescriptions for curing these eye diseases (Schering Corporation 1955). Interestingly, apart from transmitting a range of diseases to humans, as outlined below, the ancient Egyptians had recourse to use the fly itself in medicinal recipes. Fly dung mixed with sycamore juice was applied to a tumor (P. Ebers 570). Fly dung and flax boiled in oil and shaped into a paste was applied to the scalp to control hair loss

(P. Ebers 774) and as we shall see below a child's suffering was thought to be soothed by the ingestion of fly-dung.

A micro-organism, Chlamydia trachomatis, can be transmitted by flies when they alight on the eyes—resulting in the eye infection trachoma, a chronic granulomatous conjunctivitis. The ancient treatments for trachoma included the application of lead sulfide, yellow ochre, and red natron (P. Ebers 407). As the resulting eye discharge is easily passed from an infected person to others via the fingers, shared clothing items and so on, trachoma is associated with close-living conditions, poor sanitation, and inadequate hygiene. The eye (or eyes) become inflamed and after repeated bouts of infection the scarred eyelids turn inwards allowing the eyelashes to irritate the eyeball—if this very painful situation, termed trichiasis, continues untreated then blindness is the likely outcome. Remedies in the Ebers Papyrus, nos. 424 and 425 for example, are concerned with inwardly growing eyelashes termed "curling up of the hair in the eye." The recommended "cure" involved the blood from various creatures: bat, lizard, ox, ass, pig, and goat. Interestingly, in modern times, a species of Chlamydia was connected to a disease of sheep which not only caused ewes to abort but also affected pregnant women living near farms, and also caused eye disease in humans (Brothwell 1991: 18).

It has been suggested that the blind were very numerous in ancient Egypt (Ghalioungui 1963: 131; Ebeid 1999: 153) but actual evidence of eye disease from ancient human remains is unlikely to be detected as the mummification process dried out and distorted the skin around the orbits and also, in most cases, the eyeball did not survive. There are, however, other types of evidence for blindness in ancient Egypt.

Texts reveal that the term "darkness" is used, in various permutations, to denote blindness but, in some examples, caution is needed concerning interpretation. On British Museum stela no. 374, a scribe named Nekhtamun (of the Deir el-Medina tomb-building workforce) asks Meretseger, goddess of the region, who has "causest me to see darkness by day" for mercy—that is for the return of his eyesight. Turin stela no. 318 gives us the words of a sculptor, Neferronpet: "Thou causest me to see a darkness of thy making ... for health and life are in thine hand ...." The blindness referred to in these two examples (and others) may have involved disease through fly-inflicted infections but, equally, may be due to adverse working conditions, such as stone dust affecting the eyes. It should also be noted that the appeals for restored sight might be viewed alternatively as asking for religious clarity and enlightenment (Gunn 1916: 85–6), as opposed to healing a physical affliction.

In the section on bees, we have already read of the plight of the blind draftsman Pay (Wente 1990: 142), and here, in terms of interpretation, we are on a firmer footing. As stated previously, we don't know the actual cause of Pay's blindness—it could have been caused by fly-borne infection or a traumatic event or even have been age-related—but we feel his distress at the loss of his eyesight and his need for soothing medication. In the medical papyri we read of other medicants to sooth diseased eyes. A mixture of a ground green mineral and water was applied to the eyelids to cool the eyes (Ramesseum Papyrus A, no. 245) whilst the same papyrus (nos. 25–6) suggests an eye wash comprised of hemp and ground celery.

An inscription, written during the Nineteenth Dynasty, on the mummy board of Inerferti, wife of the Deir el-Medina craftsman Sennedjem (Theban Tomb no. 1) tells us that the lady is blind. She blames her misfortune on evil words said by neighboring village women, however, and more likely, she may have contracted one of the many eye diseases common in Egypt. Many of the harpists depicted in Egyptian tomb paintings have often been said to be blind (Nunn 1996: 198; Halioua and Ziskind 2005: 111). Interestingly, Ghalioungui (1963: 131) states that modern Koran or Coptic singers of hymns are often chosen from amongst the blind.

As noted above, flies could be responsible for stomach and other internal disorders—via their walking through and dissemination of fecal material. Through the fly's contamination of food, dysentery—an infection of the intestinal tract—was highly likely to have been contracted with the symptoms including severe diarrhea often containing blood and mucus. There was a high infant mortality in ancient Egypt, and an individual was open to the greatest health risks during infancy and early childhood; the problem of the survival of children's bones for examination has been summarized by Filer (1995: 22–3). Diarrhea, vomiting, convulsions, and fever—through fly contamination—was probably responsible for the death of many infants under two years of age, as it is in some modern countries (Haffejee and Moosa 1985: 1867). The time when an infant was weaned, from breast milk to a more solid diet, was also fraught with the danger of food and uncovered milk being contaminated by flies. The suffering of affected children may be detected in P. Ebers 782, in a remedy to stop a child crying—this involved scraping fly dung off a wall and mixing it with (possibly poppy) seeds, the resulting paste was strained and the child was to ingest the mixture for four days; unfortunately, if the child had dehydrating diarrhea, he/she was unlikely to have survived for those four days. Surely an ironic remedy, as a fly (or flies) was likely to have been the causative agent of the child's illness and distress in the first place.

Typhoid and cholera can also be spread by flies due to their fascination with human fecal material. Cholera, contracted from food (for example vegetables and sea food) or drinking water contaminated by infected fecal matter, results in dehydration with the imbalance of bodily fluids potentially causing death within twenty-four hours. There may be abdominal pain, vomiting, and diarrhea. Typhoid, transmitted from food and water infected by feces (Hempel

2018: 111), would have caused a high fever, rose-colored spots on the torso, sweating, chills, and possibly an inflamed spleen and liver. As the typhoid bacillus cannot survive desiccation, fecal material offered a ready source of moisture, as did the moist lips of the Egyptians, thus the disease was easily spread. There does not appear to be any easily identifiable references to these two diseases in the medical papyri and, of course by their very nature of expression, they would not leave any markers on the bony skeleton, but it is difficult to believe that they did not exist in ancient Egypt. Although not referring to these two diseases directly, the many recipes in the medical texts for ailments such as fever, abdominal pain, vomiting, and diarrhea might be referring to the associated symptoms of typhoid and cholera.

The evidence for the domestication of animals is well attested from ancient Egypt in terms of animal remains (mummified and skeletal) and artwork (Figure 2.3).

Relevant to this section on flies and to the later section on malaria, tomb paintings show men herding cattle in different environments, and wooden models (such as that of a cattle census from the Eleventh Dynasty Tomb of

FIGURE 2.3 Man milking spotted cow, Tomb of Kensit, *c.* 2000 BCE. Courtesy of Wellcome Images.

FIGURE 2.4 Fly-whisk being used, stele from the Tomb of Amenemhat and Yatu. Middle Kingdom, late Twelfth Dynasty–early Thirteenth Dynasty, 1870–1770 BCE. The Art Institute of Chicago.

Meketre, for example) indicate the close association between the ancient Egyptians and their animals (Figure 2.3). Herodotus (Book II 1988: 143) noted that "They [the Egyptians] live with their animals – unlike the rest of the world – who live apart from them." A consequence of this close relationship, unfortunately, would have been a proliferation of flies attracted to the animals' bodily warmth and excretions. The persistent behavior of flies was noted earlier and is further suggested by the use of fly-whisks. An important part of the ancient Egyptian royal regalia—the flail—was possibly derived from the fly-whisk (Figure 2.4) (Evans and Weinstein 2021: 48) and on a more basic, everyday level fly-whisks (and fans) were necessary items to ward off irritating flying insects (Figure 2.4).

In ancient Egypt the majority of domestic buildings were constructed from mud bricks—a ready source of the material was provided by the river Nile silts. As in modern times, sometimes the ancient Egyptians added animal dung (and straw) to their mud bricks to give them substance (Figure 2.5)—this encouraged flies, attracted them to the fecal matter, to enter homes and spread disease to the occupants by alighting on them or by contaminating foodstuffs.

Unfortunately, despite the valuable contribution animals made to life, they brought serious health problems to those who worked with them. Two diseases that should be considered in this context, and in relation to flies, are tuberculosis and anthrax.

Tuberculosis, in contrast to the diseases mentioned above, often does leave telltale evidence on the bony skeleton and we do have evidence for this disease from ancient Egypt and Nubia; whilst tuberculosis (caused by the bacterium Mycobacterium tuberculosis) can affect many parts of the skeleton depending on the type of bacillus involved (Roberts and Manchester 1995: 137), spinal involvement in tuberculosis (termed Potts's disease) is very characteristic and has been found in a number of skeletons and mummies (examples of which are given by Strouhal 1991), and again, flies may have been involved in the transmission of the disease. A useful review and re-evaluation of suggested cases of tuberculosis from ancient Egypt and Nubia is provided by Buikstra et al. (1993: 38–45). Tuberculosis can be spread to humans through milk and meat infected by the bovine form of the disease (Strouhal 1991: 191; Filer 1995: 18; Roberts and Manchester 1995: 136) and it can be carried in airborne droplets when an infected animal or person sneezes; the ancient Egyptians' close contact with their animals may have facilitated the spread of tuberculosis (Figure 2.3) and other diseases.

Flies can transfer tuberculosis by wading through the bodily secretions (sputum, sweat, and so on) of infected animals and then settling on the mouths and food of humans; in this scenario children would be especially vulnerable. Initially, the infection enters the lungs where the bacteria multiplies and

FIGURE 2.5 Mud bricks and building. © Joyce M. Filer.

from there it may spread to the lymph nodes or other parts of the body. Further examples of tuberculosis in human remains and possible artistic representations of the disease are discussed by Filer (1995: 29–30, 67–72), however, the medicals texts do not seem to refer directly to the disease (Nunn 1996: 74), although the recipes for lung conditions and coughs in various papyri may be associated with the symptoms of tuberculosis. Kamal (1975: 681), however, suggests that Ebers 860 and 861 are describing glandular tuberculosis.

Anthrax is an acute infectious disease of domesticated animals and is caused by the bacterium Bacillus anthracis. Humans contract the disease mostly through contact with the animals themselves and by handling their hides (Roberts and Manchester 1995: 110), hair, or excrement. Anthrax spores may be inhaled from animal fibers causing the pulmonary form of the disease. Flies alighting on infected animals or their products, however, could transfer anthrax spores to cuts or sores on the skin of farm workers resulting in cutaneous anthrax; here large blisters occurring around the sites of entry progress into black scabs, and, if left untreated, the infection may invade the bloodstream and lymph nodes, leading to death. As with tuberculosis, there are no definite references to anthrax in the medical texts of Egypt, although references to "a swelling of Khonsu" in the Ebers Papyrus, and suggested to be variously cancer, cutaneous leprosy, bubonic plague, or neurofibromatosis by Nunn (1996: 167–8), might equally be signs of anthrax infection. It is suggested that anthrax certainly existed in the Middle East by the time of the physician Avicenna (980–1037 CE) who describes the disease (Kamal 1975: 60).

In the section on mosquitoes, the transmission of filarial worms by mosquitoes will be described, but it is important to note that another species of fly—*Simulium* or blackflies—can also infect humans with the filarial worm *Onchocerca volvulus*, and these are also linked to eye disease. Through the bite of an infected female blackfly, immature parasitic worms enter the body, the developed adult then lodges in nodules under the skin and then releases more immature worms. Onchocerciasis, often known as “river-blindness” (Burgess and Cowan 1993: 46) is associated with fast flowing rivers, the blackflies’ preferred habitat, and, apart from elephantiasis (noted below in the section on mosquitoes and filariasis), other resulting conditions can be visual defects including blindness when the worms migrate to the eyes. As the ancient Egyptians were in frequent contact with the waters of the River Nile, they were likely to come into contact with these blackflies on a daily basis.

MALARIA: THE MENACE TO MANKIND

Literature from ancient Egypt hints at the misery of encountering mosquitoes when working outside. In the often-called “Satire of the Trades” we are told of the reed-cutter that “mosquitoes have slain him” (Lichtheim 1973: 186). Nowadays, we know that certain mosquitoes are responsible for the transmission of malaria and it is tempting to wonder if some of the reed-cutters of ancient Egypt, and other outdoor laborers, suffered from this often-deadly disease.

Mosquitoes have been described as “the most dangerous creatures on earth for humans” (Quick 2021: 131) and malaria, a disease transmitted by some mosquitoes, is regarded as “the most important infectious disease in human history …” (Jackson 2000: 280). In the modern world malaria kills up to about two million people yearly, with children being particularly susceptible (Fenyo 2012: 170; Jones 2012: 8). Parkinson (1991: 129–30) describes a New Kingdom copy of a Middle Kingdom spell to be said over an amuletic “knot” to protect a child from fever—whether the fever is that of malaria (or of some childhood disease) is unknown but it is worth considering as such. Malaria is still regarded as one of the world’s most dangerous diseases (Montell and Zwiebel 2016: 294), with 90 percent of current world deaths, as a result of malaria, occurring in Africa (Dobson 2007: 177) where children are particularly at risk of death (Quick 2021: 131). Malaria is probably a very old disease (Kiple 1993: 363; Dobson 2007: 177; Sherman 2007; Hempel 2018: 122); it has been suggested that environmental conditions for malaria-carrying mosquitoes were better in ancient times than nowadays and so malaria is likely to have existed from at least pharaonic times (Nunn 1996: 73; Nunn and Tapp 2000: 151) and biological research, see below, now suggests an early date for the existence of malaria in ancient Egypt and Nubia. Thus, it would be useful to consider what evidence there might be for malaria from ancient Egypt, together with some discussion on the possibility that people living in malarial areas might have had a form of genetic protection against the disease; the problems of identifying the disease in human remains will be addressed below.

There are three factors necessary for the transmission of malaria: a human host, a parasite, and a vector (i.e., a mosquito) (Stewart 2012: 318). Malaria is carried and transmitted by mosquitoes infected by the Plasmodium parasite, in particular infected females of the Anopheles species (Humphreys 1997: 98). Male and female mosquitoes subsist on plant nectar (Harrison 1978: 143) but the female also requires a blood meal in order for her eggs to develop (Harrison 1978: 144; Jones 2012: 31). Various vertebrate mammals may provide such blood meals (as unwitting hosts) but humans are the hosts of choice as human skin is more easily pierced by the mosquito's proboscis in the search for blood sustenance. The plasmodium parasites are injected into the host as blood is withdrawn (Jones 2012: 94). Should a female *Anopheles* mosquito be infected by up-taking blood from an infected (human) host then this infection is introduced into another host's system as blood is withdrawn during a feeding session—malaria then develops in the "new" human host. In turn, any uninfected mosquito then sucking blood from an infected human will also, in turn, become a carrier of malaria—and so the disease spreads. Harrison (1978: 144) stresses that mosquitoes "have firm preferences and keep quite regular hours" in terms of their preferred environment and attacking habits. Harrison (1978: 152) also outlines the broad range of water types and environments preferred by mosquitoes but Jones (2012: 56) notes that whilst they might lay their eggs in all types of water, they tend to prefer stagnant, often fetid, pools and swamps. Worryingly, the bite of only one mosquito may introduce sufficient sporozoites to establish malaria (Maegraith 1948: 1) and a person may be infected with more than one type of malaria and also be affected at different times (Dunn 1993: 859).

There are a number of species of malaria parasites that can affect humans (each distinguishable from one another under a microscope: Wilkins 1994: 176) and each producing varying degrees of fever, hot and cold sweats or chills, malaise, headaches, and intense thirst. These are *Plasmodium vivax*, the most common type causing benign tertian malaria (Wilson 1993: 384), *Plasmodium ovale, Plasmodium malariae* (a milder disease), and *Plasmodium falciparum*.

P. falciparum is the most dangerous and lethal type of malaria (Harrison 1978: 109) and the type most pertinent to this discussion of malaria in ancient

Egypt; as noted above, it is important to stress that an individual may be infected by different and/or double types of malaria at the same time. *P. falciparum* causes malignant tertian fever and has a relatively short incubation period and progresses rapidly (which I believe is an important factor in my discussion of Alexander the Great below). The most serious form of falciparum malaria is cerebral malaria (Jackson 2000: 281) and this will also be pertinent to my comments on Alexander. Following the bite of the female Anopheles mosquito (of which there are perhaps some sixty or so species capable of transmitting malaria (Brookesmith 1997: 102; Burfield and Reekie 2005: 31)) and each preferring particular types of water or environment (Dobson 2007: 178), the malarial parasites are introduced into the human bloodstream, enter the liver cells, and then invade the red blood cells, where they multiply periodically (depending on the species involved) and thus periodic fevers, characteristic of the disease, subsequently occur. In addition to fever, other symptoms of the disease may include headache, anemia, even possibly convulsions and coma (Stewart 2012: 318); the spleen of a malaria victim may become enlarged. Ebeid (1999: 208, fig. 9.26) shows a hypertrophied spleen from a Coptic mummy and suggests its gross enlargement was due to malaria. Ebeid (1999: 208–9) notes a medical case (No. 204: 67) from the Papyrus Ebers which is thought to be describing an enlargement of the spleen though here Ebeid, quite rightly, notes that malaria is only one of the causes of such a condition (such as schistosomiasis).

As noted in the introduction, some cultures have regarded "the wrath of God" as the cause of these malarial fevers (Wilkins 1994: 177) but we have no indication as to how the ancient Egyptians regarded the disease. The Hippocratic writings of between 430 and 330 BCE (Lloyd 1983: 9) relate specifically to ancient Greek cultures, however, their comments on disease may have some relevance to ancient Egypt, as do the comments of the Greek author Herodotus. The Hippocratic writings, when comparing the "causes" of epilepsy and malarial fevers, state "Quotidian, tertian and quartan fevers are among other diseases no less remarkable and portentous and yet no one regards them as having a divine origin" (Lloyd et al. 1983: 237); nowadays, of course, we have identified the mosquito, via its infected bite, as the causative culprit of malarial fevers. Although modern tests for malaria (outlined below) suggest the disease was present during various periods of Egyptian history, Nunn (1996: 73) reminds us that "the [Egyptian] medical papyri ... are silent on the characteristic recurrent fever at three- or four-day intervals, which surely could not have escaped notice." Perhaps such fevers were a commonplace event in Egypt and were thus not worthy of comment? We should also take on board the fact that most adults living in endemic malarial areas might have a lifelong exposure to malaria (Stewart 2012: 318), thus their symptoms may be mild to moderate and only for a short time. Were malarial fevers perhaps commonplace?—this may have been the situation in ancient Egypt.

Again, as noted in the Introduction to this chapter, the development of agriculture went hand in hand with the evolution of human diseases, which would become endemic. Civilizations that depended upon irrigation systems to water farming land were particularly at risk of attracting mosquitoes and, of course, Egypt is a prime example of such irrigation-based farming.

Following the hoped-for successful inundation of the River Nile, excess water was diverted to crops further away from the river via channels cut through the land. Unfortunately, whilst crops benefitted, pools of stagnant water remained, providing excellent breeding grounds for a variety of insects. Thus, we should look to the agricultural areas of ancient Egypt for evidence of the inhabitants' experience of mosquitoes—and probably malaria. Upper Egypt, with a dry climate and large tracts of desert, had relatively narrow strips of cultivated land along the banks of the River Nile, which were watered by irrigation channels cut through the land, an environment suitable for some species of mosquito (Figure 2.6).

As the River Nile flowed, and still flows, north towards the Mediterranean Sea it spread out into a fan-like shape forming the Nile Delta—resulting in an area of tributary streams providing fertile, but somewhat marshy, cultivable

FIGURE 2.6 Irrigation scene. © Joyce M. Filer.

land; whilst this description is set in the past, much remains the same in the Egypt of today, however, the changes in the behavior of the Nile (in relation to the inundation of the past) due to the building of the Aswan Dams should be noted. Swift-flowing rivers, as was the Nile in the past, is unsuitable for mosquito larvae (unlike the blackflies mentioned above) but stagnant pools and irrigation channels, the banks of clean streams, tree holes, salty and fresh marshes are just some of the breeding places preferred by different types of mosquitoes, depending on their species (Harrison 1978: 152). Oases of fertile land, such as that in the Fayum, also existed in desiccated areas. Interestingly, during the Middle Kingdom period, King Sesostris II (*c.* 1880–1874 BCE) drained the marshes of the Fayum area "over the entire land from Memphis to the sea he dug frequent canals leading from the river ..." (Diodorus Siculus, Book I.57.3, trans. 1933)—thereby increasing land available for cultivation but, unfortunately, as a by-product, possibly creating areas favorable to some species of mosquito.

There was an early association between marshy areas and fevers (Hannaway 1993: 293, 295; Wilkins 1994: 177; Dodson 2007: 179–80; Corfield 2012: 153). In "Airs, Waters, Places" (Lloyd et al. 1983: 152) the Hippocratic author notes that the type and quality of water "plays a most important part in health" and states stagnant marsh or lake water has an unpleasant smell and those who drink such water have distended stomachs and enlarged spleens—the latter symptom often being an indicator of malaria, as noted above. Certainly, the establishment of trade routes across the Classical world helped spread malaria and it is clear that the citizens of the Aegean basin suffered from malaria as did the Romans from about the second century onwards. Importantly, it has been suggested that malaria played a major role in decimating Rome (Kiple 1993: 363) and Greece (Angel 1966: 760–3), although Exner et al. (2004: 171) describe Angel's view as "imaginative." The ancient Egyptians were also great traders—but it is difficult to assess if the spread of malaria via certain mosquitoes affected them to the same extent as it appears to have done in the Classical world.

With regard to ancient Egypt the classic reference to insects, bites, and marshy areas can be found in *The Histories* of Herodotus. Writing during the fifth century BCE Herodotus informs us that "the country [Egypt] is infested by swarms of gnats" (Herodotus, Book II, 164). Herodotus goes on to explain how the populace dealt with these annoying insects. People living in the marshlands (i.e., the northern Delta areas) affixed a net around their beds "for," as Herodotus tells us, "anyone to sleep wrapped in a cloak or linen would be useless, for the gnats would bite through them: but they do not even attempt to get through the net" (Herodotus, Book II, 164). Prior to this, Herodotus informs us that the Egyptians living in the south of the country avoid these insects—which cannot fly in the winds—by sleeping on high structures. With regard to the point about flying, Harrison (1978: 28) informs us that the mosquito is a weak flier and does not fly high which is why people sleeping above ground are unlikely to be bitten thus giving some credence to the comments of Herodotus that Upper Egyptian inhabitants slept on "towers" built higher than the prevailing winds. Interestingly, it has been suggested that a wooden frame belonging to the Fourth Dynasty queen, Hetepheres, was a support for a mosquito net (Strouhal 1991, cited in Hawass et al. 2010: 646) and, if so, indicates a very early date for the presence of potentially malaria-carrying mosquitoes in ancient Egypt. Although the above reference in Herodotus suggests mosquitoes were attacking people during the night, it should be noted that some species of mosquito prefer to bite during daylight hours (Jones 2015: 163). The many Egyptian reliefs depicting herdsmen wading through waters with their cattle demonstrate the ease with which such daylight-loving mosquitoes could come into contact with people, although these are less likely to be malaria-bearing species.

It is suggested from his later comments that Herodotus himself experienced this "infestation" of insects when he visited Egypt. Modern visitors to Egypt (and Sudan) will sympathize with Herodotus (and the ancient Egyptians) regarding the annoying behavior of insects, especially mosquitoes! At this point it is important to consider the use of the word "gnats" in some translations of *The Histories*. In de Sélincourt's (1988) translation the term "gnats" is used (as Rawlinson did in 1859: 132) whereas other translations (such as Scarre 1992: 123, for example) specifically use the term "mosquitoes." What is the likelihood that Herodotus's annoying insects were mosquitoes (and therefore potentially malaria-carrying) and not merely gnats? However, the situation may not be as clear as we would like it to be. Translators are still unsure as to the exact identification of the insects mentioned in Classical texts (Jones 2012: 12) and the term "gnat" has been used very widely in literature whereas the term "mosquito" has a more modern ring to it (Jones 2012: 15). Yet, in his Commentary on Book II of Herodotus, Lloyd (1976: 382) states "gnat equates with mosquito." Interestingly, the next lines of *The Satire of the Trades* (referred to at the beginning of this section) are: "Gnats have slaughtered him, He is quite worn out" (Lichtheim 1973: 186)—clearly this translator is distinguishing between gnats and mosquitoes. Indeed, in discussions with a Greek archaeologist based in London, concerning Herodotus's use of the (Greek) word Κωνωϖες, she stated that, etymologically, it derives from Κωνωψ, meaning "mosquitoes" and that perhaps Herodotus, should he have so wished, could have used σΚνΙψ, meaning "gnat" (or plural σΚνΙϖες) had he really been referring to gnats (Mersini-Eirini Koutelou 2019, personal commuication). Thus, mosquitoes seem to be the insect intended by Herodotus.

As the modern world is well aware, nets can provide protection against mosquito attack. Importantly, if fishing nets were used in this capacity (as noted by Herodotus) then they would need to be of a very fine mesh in order to be effective. We know the ancient Egyptians were skilled at weaving linen to varying degrees of fineness, therefore, why should they not have applied the same skills to fishing nets? Lloyd (1976: 383) feels Herodotus is right concerning the use of nets. Further, it is suggested that the nets may have been smeared with a fish oil as a mosquito deterrent (Curtis et al. 1990: 8). Certainly, the Egyptians would have had access to oils processed from the many and varied types of fish they caught (Herodotus, Book II, 163–4 notes the abundance of fish available to the Egyptians), but I would also like to propose another type of oil readily available to the Egyptians and with which they might also have smeared their nets. Castor-oil, pressed or boiled from the castor plants which grew along the banks of the Nile and lakes, was used medicinally and as a lamp oil, and could also have been a useful alternative to fish oil. Interestingly, Herodotus stresses twice that the castor-oil the Egyptians made "has a disagreeable smell" and "the smell is unpleasant" (Herodotus, Book II, 164) and, as modern research informs us, mosquitoes have a very finely tuned sense of smell, in order to detect their human prey (Jones 2012: 43; Jones notes they can detect, and are attracted to, carbon dioxide in the breath) thus perhaps ancient Egyptian mosquitoes were deterred by the use of this unpleasant smelling oil on nets.

Modern research has shown that the human immune response to infection from malarial parasites is influenced by a number of factors including the (human) host's age and genetic make-up, together with the level of endemicity and epidemiological factors and the type of parasite species involved and its parasitic stage (López et al. 2010: 1). This is a complex situation—even for modern times—so we cannot hope to fully understand the impact of malaria on an ancient civilization spanning thousands of years. However, modern testing techniques can help indicate the antiquity of the disease, through the examination of ancient human remains.

Unlike some diseases (such as tuberculosis, leprosy, osteoarthritis, and so on), malaria in itself does not result in gross pathological skeletal changes on the body (Nunn 1996: 73; Humphreys 1997: 98; Hempel 2018: 122). However, modern scientific tests on samples from ancient bodies have proved interesting. The ParaSight TM-F Test detects an antigen produced by *Plasmodium falciparum* (Nunn 1996: 73) and can be detected in the skin, muscle, lungs, and brain of human remains (Nunn and Tapp 2000: 151).

This test was applied by Miller et al. (1994: 31), and cited by many authors (including Filer 1995: 8 and Nunn 1996: 73), to samples of human remains from various periods of Egyptian history including: four naturally desiccated bodies (from the Predynastic period site Gebelein), one intentionally mummified body dating to the New Kingdom (*c.* 700 BCE), and eleven Nubian Ballana (X-group) period individuals from North Argin, Nubia (350–550 CE). The parasite antigen (*Plasmodium falciparum*) was detected in samples of skin from three of the predynastic individuals, in a sample of muscle from the New Kingdom female mummy, in two samples of lungs (held in the British Museum Department of Ancient Egypt and Sudan), and in a sample of brain from one North Argin individual. The authors state that all of the seven affected individuals were suffering from "untreated malaria at the time of their deaths" (Miller et al. 1994: 31), surely the implication here being that the disease may have actually contributed to the deaths of these individuals; the underlying inference also being that these individuals possibly suffered from tertian fever having been in contact with the Anopheles mosquito. The wide range of dates (of the samples studied) supports the notion that malaria has a long history in ancient Egypt and Nubia. The positive result of malaria for the New Kingdom (*c.* 700 BCE) female mummy was reiterated by Ikram and Dodson (1998: 100); Ikram being one of the contributing authors to the 1994 Miller report. There seems, however, to have been concerns when other researchers attempted to replicate these results with specific regard to this New Kingdom mummy. Currently housed in the British Museum, this particular female mummy, named Irtyersenu, is also known as the Granville mummy having been the subject of both an autopsy by Dr. Augustus Granville in the early 1800s (in the company of a number of scientific professionals) and a publication of the findings (Granville 1825). However, when further samples (from leg muscles) from this mummy were tested they showed negative results for *P. falciparum* (Taylor et al. 1997, cited in Bianucci et al. 2008: 1881). In a later, and most interesting, paper describing the life and achievements of Granville, Harer (also a contributor to the Miller 1994 report) and his co-author do not mention malaria but do suggest that Irtyersenu (who died at *c.* fifty years of age) had pulmonary tuberculosis and thus her death was probably due to respiratory failure (Harer and Tapp 2014: 105, 106); thus, the malarial status of the mummy Irtyersenu remains debatable.

Further evidence of active malaria when an individual was alive has been suggested by examinations of other ancient Egyptian human remains, some of which will be presented here. ROM I (housed in the Royal Ontario Museum), the mummy of an adolescent male named Nakht, a weaver, was autopsied and also tested positive for malaria (Cockburn et al. 1998: 103). Interestingly, in a letter to *The New England Journal of Medicine*, Boston (1981), the Cockburns, following a correction of a reviewer's comment on their book (when originally published in 1980), also revealed that they had considered Nakht's death to have been due to splenic rupture (on observations made during the autopsy) but that this diagnosis was left out of the official report as there was disagreement amongst the researchers involved in the project. However, as already noted, damage to the spleen is an important consideration in a diagnosis of malaria and the Cockburns further state that a sample of Nakht's spleen was forwarded to the U.S. Public Health Service Hospital in New Orleans where the tissue tested "strongly positive for malaria by the immunofluorescent-antibody test" (Cockburn and Cockburn 1981: 1534). According to this letter these findings were later debated upon, and accepted as genuine, by the Royal Society of Tropical Medicine and Hygiene in November 1981 (Cockburn and Cockburn 1981: 1534).

Artemidorus, the mummy of a young adult male, currently housed in the British Museum (catalogued as EA 21810) and dating to 100–20 CE, was CT-scanned and reported upon by Filer (1998: 21–4). Evidence from the mummy's beautifully decorated stucco case indicates that the mummy of Artemidorus provides an interesting example of the meeting of various cultural influences: a Greek personal name, a Romanized portrait, and traditional Egyptian funerary practices. The scans revealed traumatic damage to the nose and the back of the skull which, showing no signs of healing, may have been the cause of death or alternatively as the result of rough handling by the embalmers in ancient times—certainly we are looking at a peri-mortem event. During the examination of this mummy, I had cause to suspect that Artemidorus may have had malaria when alive; this suggestion will be reported upon at a later date. Certainly, the mummy's Greek name and the Roman dating of the mummy package might be important in the light of the possibility that malaria seriously decimated parts of the Classical world (as noted above).

Nerlich and colleagues studied ninety-one tissue samples from ancient Egyptian skeletons and mummies dating from 3,500 to 500 BCE using the techniques of DNA amplification and gene sequencing. In a 2008 online interview Nerlich reported the identification of ancient DNA for the malaria parasite *P. falciparum* in tissue samples from two Theban mummies.

The mummy of an Egyptian child (aged fifteen to eighteen months at death) and dating to the end of the Early Dynastic period or the beginning of the Old Kingdom period, was CT-scanned before immunological investigations were performed to determine whether or not malaria had caused the child's death (Bianucci et al. 2008: 1881). Samples from the child mummy tested positive for *Plasmodium falciparum*, but were negative for mixed *P. falciparum-P.vivax* infections which is an important finding in the light of my earlier comment that people can be infected with more than one type of malaria (Dunn 1993: 859), or a double malarial infection as indicated in the new findings for Tutankhamun discussed below.

Two well-known individuals from ancient Egypt are also involved in our story of malarial infection: Alexander the Great and, as just noted above, Tutankhamun.

It was mentioned above that *P. falciparum* malaria may progress into the severe symptoms of convulsions, coma, even death and these indications may be important when regarding the case of Alexander the Great. Over the years a relatively large number of causes of death have been proposed for Alexander: poisoning (possibly deliberately by strychnine or accidentally through lead or methanol in his wine), alcoholism (leading to hepatitis), leukemia, typhus, West

Nile fever, typhoid fever, and complications of a lung perforation (possibly initiating pleurisy) following an arrow wound to the chest (for an overview see Chugg 2004: 16–20; Saunders 2006: 28–9); however, a perusal of available evidence (i.e., ancient texts) would not fully support any of these suggested diagnoses. Saunders (2006: 28) sounds a warning concerning the reliability of the available ancient texts in that they were written quite some time after Alexander's death and that "not one is free of contamination, retrospective propaganda, self-serving lies, half-truths." Malaria, first suggested during the modern nineteenth century as the cause of Alexander's death, however, seems to fit all the features of Alexander's final illness and yet, it has been noted, whilst modern authors have "endorsed" the idea of Alexander's having suffered from malaria "it never seems to have been put forward with quite as much force and detail as it merits" (Chugg 2004: 21), thus it would be useful for us to examine, briefly, some of the ancient textual sources for possible evidence of Alexander's (and at least one of his companions) having contracted malaria.

In the autumn of 324 BCE (Chugg 2004: xxiv, 6), we are told, Hephaestion (one of Alexander's soldiers and his closest friend) caught a fever (Plutarch, trans. 1985: 329) with Alexander receiving a message to the effect "that Hephaestion's condition was serious" (Arrian, trans. 1986: 371) but died before Alexander could visit his friend. It seems Hephaestion was ill for at least seven days before he died (Arrian 1986: 371). As his condition progressed towards death, seemingly within two weeks, it is possible he had succumbed to falciparum malaria, however, Saunders (2006: 20) suggests typhus as an alternative disease. At this point, two further comments concerning illness should be noted: that some of Alexander's (and Haphaestion's) soldiers contracted a skin disease after entering the waters of a salt lake (Quintus Curtius Rufus, trans. 1984: 235) which was easily spread to their comrades (but is unlikely to be connected to malaria) and that, following a meeting with coastal Indians, who would not trade with the Macedonians resulting in the latter's suffering from famine and starvation, the Macedonians were beset by a hitherto unidentified "plague" which killed many of them (Quintus Curtius Rufus 1984: 236)—again, this may or may not be connected with malaria, but it is worth noting. Certainly, some two weeks before his final illness Alexander had sailed in swampy malaria-infested areas. He had decided to improve the Babylonian Pallacopas canal(s), originally built by the Assyrian kings, which, according to Strabo 16.9 (trans. 1930: 205) "easily filled, and their mouths choked, by the silt" providing ideal breeding grounds for mosquitoes. Strabo 17.I.17 (trans. 1932: 31), in comparing Alexandria in Egypt with other cities, tells us that "other cities that are situated on lakes have heavy and stifling air in the heats of summer, because the lakes then become marshy along their edges ... when so much filth-laden moisture rises, the air inhaled is noisome and starts pestilential diseases." Indeed, it is noted that as Alexander was returning to

Babylon via the marshlands his ship was delayed (Diodorus of Sicily, 1 7.116.4, trans. 1963: 465) and "menacing clouds of mosquitoes whined about in the sultry air" (Chugg 2004: 10)—thus surely the scene was set for Alexander's contracting mosquito-borne malaria? It is also important to note that adults living in malarial areas, having accustomed exposure to mosquitoes and malaria, are likely to experience only mild to moderate symptoms and are less likely to succumb fatally to the disease (Fenyo 2012: 170; Stewart 2012: 318) whereas newcomers to an area encountering a disease for the first time, such as Alexander (and his companions), could succumb to a newly encountered disease quite easily.

Perusal of Classical texts indicates that Alexander died in June 323 BCE after *c.* ten days of illness (Plutarch, trans. 1985: 332). Although it should be noted that there can be extreme variability in the symptoms an infected person may exhibit, the progress of Alexander's illness would suggest falciparum malaria (i.e., malignant malaria), and most likely, the cerebral form, the most serious consequence of the disease; Maegraith (1948: 18) describes in detail the various and disturbing symptoms of cerebral malaria. The falciparum parasite has a particular effect on the blood cells—making them sticky and clumping together, depriving the organ they are invading of oxygen and thereby killing it. This frequently happens in the brain and so the sufferer enters into a coma and then dies (Harrison 1978: 118). Alexander became feverish during the two weeks after traveling in a mosquito-infested area, he was lethargic and had a reduced appetite, back pain, and intermittent fever. That he may have been entering into a coma (and fatal cerebral malaria) may be extrapolated from the Classical sources when they state Alexander was "speechless and remained so" (Plutarch 333) and "He recognized his officers when they entered his room but he could no longer speak to them. From that moment until the end he uttered no word" (Arrian 393), he then lost consciousness (Saunders 2006: 28) and died.

The tomb of Tutankhamun was discovered in 1922, the king's mummified body being examined a few years later by Howard Carter and Douglas Derry (Derry 1927); there have been further reports on the king's remains (for example, Harrison and Abdallah 1972 and see Hawass et al. 2010 discussed below)—the general consensus being that Tutankhamun died aged at *c.* eighteen–nineteen years. Over the years, as in the case of Alexander the Great, quite an extensive list of possible causes of Tutankhamun's death has accrued (see Rühli and Ikram 2014 for an interesting review of suggested causes of death). Among the many proposed death scenarios, we find a fall from his chariot, septicemia, poisoning, tuberculosis and, the most "popular" and enduring theory, a blow to the back of the head. Having been privileged in being allowed to examine some of the X-rays from the 1966 Harrison examination of Tutankhamun I can state that there is no apparent head injury. The loose pieces of bone observable within the skull (see Filer 1995: frontispiece, for copy of lateral skull X-ray), and often thought to indicate a blow to the head, have now been clearly shown to be sections of vertebrae (Boyer et al. 2003: 1146, figs. 1–3) probably displaced during the original Carter and Derry examination. If these pieces had been broken off during the mummification process, then we would expect the broken pieces to be embedded in the resinous layer(s) within the skull, which they are not. In my opinion, there is no indication of cranial assault by a murderer! Even though some of the results have been questioned (Pays 2010a, 2010b; Timmann and Meyer 2010), the latest examination of Tutankhamun's mummy has produced some interesting findings (Hawass et al. 2010), of relevance to this discussion on malaria.

In addition to this new examination of Tutankhamun (Hawass et al. 2010), the bodies of ten other ancient Egyptian individuals, thought to be related to Tutankhamun, were also examined during this project (plus five mummies from a different time period as a control group). The bodies were CT-scanned and samples—for possible genetic/familial information and disease detection—were taken and analyzed. Within the scope of this chapter the overall genetic/familial and disease results for these bodies will not be discussed except for those pertaining specifically to this discussion of malaria.

The Hawass et al. (2010) results for Tutankhamun revealed not only the identification of DNA of the malaria parasite—*Plasmodium falciparum*—indicating the king had been suffering from malaria at the time of his death but were "indicative of at least a double infection with the P. falciparum parasite" (Hawass et al. 2010: 643, 644), termed "malaria tropica" in the report. It is assumed to have contributed to the king's death (Hawass 2010: Table 3). The same results—multiple infections of malaria—were also found for the mummy of Yuya (Hawass et al. 2010: 643, 644), Tutankhamun's grandfather or great grandfather (Hawass et al. 2010: 640, Table 1) in his maternal lineage. Thuya, Tutankhamun's grandmother or great-grandmother (Hawass et al. 2010: 640, Table 1), also tested positive for one strain of malaria (Hawass et al. 2010: 644). Reiterating the authors' comment that we do not know if the disease caused the deaths of the individuals testing positive for malaria in this investigation (Hawass et al. 2010: 646), the fact that they had the most dangerous form of malaria (*P. falciparum*) is significant. Of course, the response of an individual to an infection is affected by that person's particular constitution. Both Yuya and Thuya were at least fifty years of age at death and so are either likely to have contracted malaria late in life or had built up a partial immunity through repeated exposure to malaria pathogens during their relatively long lives. However, further results from the Hawass et al. project suggest that Tutankhamun was not in the best of health, apart from contracting malaria, and that his immune system may have been compromised. The king appears to have had malformations in both feet, possibly necessitating the aid of walking sticks and may have suffered a leg fracture. Thus, being in a weaker state, he could have been more susceptible to the often-severe symptoms of falciparum malaria.

Ancient Egypt provides pictorial representations of their kings (and members of the royal family) engaged in daily pursuits—both in religious and less formal contexts. Whilst we do not know how far these represent the actual truth it is worth noting some particular artistic evidence for Tutankhamun's "domestic" life which might be relevant to his "modern" diagnosis of malaria. Scenes decorating a wooden shrine covered with sheet gold and standing on a silver-overlaid sledge (A Guide to the Egyptian Museum, Cairo, 1986, item no. 14, p. 271) show Tutankhamun and his queen, Ankhesenamun, engaged in sporting activities—one scene shows the king shooting birds with a bow and arrow. That here the king is seated for this activity, for which ancient Egyptians are normally depicted standing, has been suggested as evidence of his infirmity (Hawass et al. 2010: 645); another scene on the shrine shows the couple hunting fowl in a papyrus skiff, also in the marshlands. That these scenes take place in marshy areas might indicate the ease with which Tutankhamun, and also the populace in general, might come into contact with marsh-loving, malaria-carrying mosquitoes.

The mummies of two female babies were found in the Treasury of Tutamkhamun's tomb (KV 62): one (no. 317a) was a fetus "not exceeding 5 months," the other (no. 317b) was perhaps of seven-months gestation (Leek 1972: 22, 23). Hawass et al. (2011: W830, W31, Table 2) suggest 24.7 weeks (5–6 months) and 36.7 weeks respectively. They are thought to be the daughters of Tutankhamun (and his wife Ankhesenamun) and, as currently we know of no other potential heirs in direct line from Tutankhamun, so these two girls might be regarded as the end of the line in terms of the royal succession. That both fetuses died prematurely and that it has been proposed (but debated) that the "elder" had skeletal deformities (Filer 1995: 61, figs. 36 and 37 following Harrison et al. 1979: 19, 20) and also that their putative father Tutankhamun also seems to have had skeletal (foot) malformations, is it possible to suggest, tentatively, that there were inherited "medical" problems in the family? Skeletal deformities in the elder fetus, however, were not found by Hawass et al. (2011: W835, W836). Clearly there were problems as the two babies were unable to develop to full term, although it is suggested that the larger fetus (no. 317b) was possibly stillborn at eight–nine months (Harrison et al. 1979: 20). Usually regarded as separate (in time) births, interestingly, Chamberlain (2001: 9) suggested the babies were twins and that they had a condition termed Twin-to-Twin Transfusion wherein one twin grows larger at the biological expense of its sibling and thus this accounted for the size and seeming age discrepancies between the two girls—this was reviewed and discussed by Hellier and Connolly (2009: 168–70).

Added to this, might it not be possible to propose that the often-fatal form of malaria, *P. falciparum*, had a part to play in the end of this particular ancient Egyptian family and dynasty by "killing off" weaker individuals? As stated, Tutankhamun seems to have been in a weakened state of health and, even in endemic malaria, infants (and children) are unlikely to have immunity to the disease (Dunn 1993: 858) and so perhaps the fate of these babies, seemingly the last of their line, was sealed by this deadly disease. The Hawass et al. (2010) report does not indicate malaria for the two fetuses, neither is there a clear identification of the mummy of their mother. Thus, in the absence of further research so far, it might be useful to consider that had their mother also been infected by malaria, especially during her pregnancy, then this could have compromised the fetuses (either separately or as twins, if indeed they were) in utero. In congenital malaria, the parasites are passed on from mother to child before or during birth (Brookesmith 1997: 102). Other modern research has shown that women are more susceptible to malaria during pregnancy (Bray and Anderson 1979: 427, 429) and that *P. falciparum* can cause "substantial placental damage" (Taha et al. 1993: 319) sometimes resulting in retarded intrauterine growth (Taha et al. 1993: 319). Conroy et al. (2019: 553) reported on a pair of modern East Ugandan twins, one of which (Twin B) had, at one point, contracted a placental malaria infection. In follow-up sessions one year after birth, both twins were found to be delayed in various functions (e.g., sitting, standing, speaking) with Twin B attaining much lower scores across all areas of function performing (Conroy et al. 2019: 553). These modern reports into the effects of malaria on fetal development illustrate the serious effects of the disease in utero and may have some relevance to the deaths of the two little girls found in Tutankhamun's tomb; further, perhaps we should reconsider the interesting reports of Chamberlain (2001) and Hellier and Connolly (2009).

Certainly, the testing of mummies for malaria offers an encouraging way forward in detecting the history of malaria in ancient Egypt and Nubia and how it might have affected society. This writer, for one, looks forward to further work in this field.

Although we are still far from understanding the effective role played by malaria in past civilizations (Bianucci et al. 2008: 1880), the above information would suggest that malaria does have a long history in ancient Egypt and Nubia, but how did the ancient Egyptians deal with mosquitoes? As we have seen from Herodotus at least two methods of control—the use of nets (perhaps smeared with an oil) to sleep under and sleeping on high structures—were undertaken in ancient Egypt. Ebeid (1999: 208 and note 64) citing Ghaliounghui, notes the use of mosquito repellents in the Papyrus Ebers (No. 846: 214) in which the "sufferer" is advised to smear on behen (possibly balanites)-oil; the fruit of the balanites tree is frequently found in archaeological contexts but its ancient name is not confirmed (Manniche 1989: 81). Herbs have been utilized throughout human history as insect repellents (Corkhill 2011: 20). As in ancient Greece and Rome, it is possible that rosemary was also burnt in ancient Egypt as a fumigant against insects (Corkhill 2011: 42), however, despite its being grown in modern Egyptian gardens its cultivation in ancient Egypt is uncertain (Manniche 1989: 144). In modern times a malarial remedy is made from Artemisia, the sweet wormwood plant, and thus may have been the case in ancient Egypt. Manniche (1989: 80) citing the works of Prosper Alpini, a physician who traveled in Egypt during the years 1581–1584, noted the use of Artemisia to "expel a bilious fever" and its use as a massage oil "to bring out excess body heat before an attack of contagious fever"—perhaps this medieval usage had its roots in ancient times? Cinnamon has been used, in traditional practices, to ward off mosquitoes the wood of which could be burnt as a fumigant (Manniche 1989: 90). Harrison (1978: 209) relates how in modern times smoke fumigation has been utilized in homes to kill or stun mosquitoes. In the medical papyri we read of recipes for fumigants to rid the home of insects—perhaps these, together with the smoking method still used in modern times, were employed in ancient times.

MALARIA AND ANEMIA: GENETIC PROTECTION?

There has been much research considering the presence of anemia(s) in archaeological populations. It has been suggested that, in modern times, people living in malaria-affected areas may have a genetic anemias, a protection against malarial disease, and that this may have been the case in antiquity (Angel 1966; Humphreys 1997: 98). Certain markers, pitting in the eye sockets (termed cribra orbitalia) and pitting on the skull (termed porotic hyperostosis), have been suggested as indications of anemias. Whilst these markers are often thought to be the result of iron-deficiency anemia it is also considered that such genetic traits offer resistance to malaria. Such markers may indicate that malarial disease has been present in enough generations of the given population for these genetic traits to be augmented through natural selection.

The work by Stuart-Macadam suggests that anemia results in these pitting episodes (see summary in Stuart-Macadam 1991). Over the years many causative factors for these lesions have been proposed including: infectious diseases, parasitic infections leading to blood loss, the effects of weanling diarrhea, and an iron-deficient diet (Larsen 1999: 33–4). Once it was thought, negatively, that porotic hyperostosis (and cribra orbitalia) indicated that individuals (and thus populations in a wider context) were under the stress of poor health and so were unsuccessful in adapting to their situation. Now, however, research suggests a contrary position, "that it is actually a positive response and a sign of a healthy defense system" (Stuart-Macadam 1991: 37), whereby microbial parasites are less attracted to hosts with lowered iron levels and hence blood of a "lesser quality" and so it may be possible to suggest that the loss of certain nutrients, iron being a case in point, could actually benefit people. With their keen sense of smell, it is likely that mosquitoes can detect the quality of a potential host's blood—thus, blood lacking nutrients, or having malformed blood cells (through perhaps sickle-cell anemia), would be both unattractive and non-beneficial to these blood-seeking insects. Having had personal experience of such a situation, with a platelet disorder, I can testify that I am not high on the mosquito dining menu, in fact, mosquitoes avoid me! Therefore, is it possible that many lower-class ancient Egyptians, possibly due to an iron-deficient diet, were unattractive to mosquitoes? So far, we have most evidence of malaria (active in life) from the mummies of the wealthier people in Egyptian society, the ones who had access to a better and probable iron-rich diet.

Concerning a genetic component at work it has been suggested that the occurrence of sickle-cell anemia (and the thalassaemias) in populations living in malarious regions might indicate a form of genetic protection. Whilst full sickle-cell anemia is a debilitating and often fatal condition, a partial sickling trait would seem to confer some measure of immunity from the often-severe symptoms of the most dangerous form of malaria—*P. falciparum* malaria. Whilst we do not know the full situation from ancient Egypt and Nubia, modern evidence reveals that the people of the Nile Valley are subject to four main anaemias: sickle-cell anaemia, β thalassaemia, sickle-cell/β thalassaemia (a combination of the first two), and chronic iron-deficiency anemia (Hillson 1980: 173). We need more research into the types of anemias which occurred in ancient times.

Certainly, there is evidence of cribra orbitalia and porotic hyperostosis in ancient Egyptian and Nubian populations. An interesting individual example of fairly severe porotic hyperostosis occurring on a skull is that seen on an Eleventh Dynasty female skull currently housed in the British Museum, London. Years ago, cross-sectional samples were removed from this skull demonstrating the depth and involvement of the observed condition. That this skull is suggested to be that of a "princess" or high-ranking individual raises questions about her situation. Being supposedly "royal" or noble we would expect her to have access to meat (food of the privileged) and therefore not be iron-deficient; however, there are also the possibilities that the lesions indicate she had a genetic immunity (to malaria) or was, in fact, seriously ill from another cause. The skeletal remains (of a formerly mummified body) of an infant, also in the British Museum, exhibits severely bowed limbs and ribs indicative of osteogenesis imperfecta or "brittle bone" disease, arising from an inadequate formation of bone tissues (Filer 1995:61). The skull has what is sometimes referred to as a "Tam O'Shanter" deformity and also shows severe "anemic" lesions on both parietal bones indicating, together with the deformed post-cranial bones, that this youngster was seriously ill before undergoing an untimely death (Filer 1995: 63, see book cover and pl. VI for illustrations of this skeleton and detail of leg bones). It is possible that this child would have had some genetic immunity to malaria but, unfortunately, congenital "brittle bone" disease took its deathly toll.

Whilst studying individuals is always interesting it is, however, useful to examine population groups for cribra orbitalia and porotic hyperostosis so that a profile of the group can be built up with regards to their dietary habits and possible disease "immunity." I have recorded many examples of cribra orbitalia (eye lesions) in skulls, for example in the Egyptian "E" Series (held in the Department of Biological Anthropology, Cambridge), dating to between 600 to 300 BCE, which, at the population level, may be indicating a form of genetic anemia, although an overall dietary influence cannot be ruled out.

Years ago, I was involved in the excavations of many tombs and graves at the multi-period (Meroitic, Post-Meroitic, and Christian periods) site of Gabati—situated at the convergence of the Nile and Atbara Rivers in Sudan. There were many examples of cribra orbitalia lesions observed and in some cases individuals, who had been buried in the same tomb, exhibited the condition—for example two juveniles, possibly siblings, were buried in the same tomb, side by side and both had noticeable lesions of unhealed cribra orbitalia, suggesting either their diet was similar or that possibly they had a genetic immunity to malaria—or both! Of course, their deaths may be attributed to other causes.

Is it possible that the observations of porotic hyperostosis (in the orbits and/or the parietal bones) in many groups of ancient Egyptian (and Nubian) skulls indicate that they had a form of genetic anemia and thus had a selective advantage over individuals without the trait and so were relatively protected from malaria?

Of course, it must be reiterated that there are also a number of non-genetic environmental factors, noted above, which might influence the occurrence of cribra orbitalia and porotic hyperostosis. It would be an interesting research project to investigate to what extent ancient Egyptian and Sudanese individuals (and other archaeological populations) exhibiting cribra orbitalia and/or porotic hyperostosis also test positive for malaria—a potentially awe-inspiring project!

MOSQUITOES AND FILARIASIS: AN ANCIENT AND DISFIGURING DISEASE

Describing filariasis as "an ancient and disfiguring disease" in 2004, the World Health Organization stated that "Egypt [in 2004] stands on the verge of eliminating lymphatic filariasis" and "begins the final chapter in stopping a dreaded disease" (WHO 2004) through the administration of drugs. The then statistics indicated that several million modern Egyptians suffer from the condition. Is stated that there is no cure for elephantiasis (lymphatic filariasis) once it is established within the body but the condition, nowadays, can be managed in order to improve the sufferer's quality of life (WHO 2004) and millions, despite drug treatment, continue to suffer from the symptoms of lymphatic filariasis (Dobson 2007: 230). Some authors suggest the condition is attested from Pharaonic times; however, we do not know to what extent the ancient Egyptians may have been affected.

Apart from malaria, mosquitoes are also responsible for transmitting filariasis, a group of diseases, the most well-known of which is elephantiasis (Wilkins 1994: 175); the "consequences have been given the name elephantiasis, such can be the severity of the oedema" (Mills 1993: 98). Filariasis is caused by tiny parasitic nematode worm larvae introduced into the lymphatic system following the bite of an infected mosquito (Mills 1993: 98; Dobson 2007: 230), however, in these cases the principal mosquitoes responsible are those of the genus Culex (Margotta 2001: 165; Quick 2021: 131), although Anopheles might transmit the disease. If left untreated the filarial worms can "block" areas of the body, thus parts of the victim's body (principally the limbs, breast, and the scrotum, but the scalp may also be affected) swell up and become enlarged, sometimes to a quite an exceptional size (for photographic examples see Wilkins 1994: fig. 175 left and right and fig.176 left and right which show examples of gross swellings of the legs, scalp, and scrotum), hence the term "elephantiasis." Nunn (1996: 71) cites Weeks for representations of enlarged male (external) genitalia in Egyptian tomb scenes.

The World Health Organization states that, today, it is the Delta region of Egypt that is primarily affected by filariasis and this may be because the area is marshier and therefore attracts more mosquitoes but, again, it is difficult to assess what the situation would have been in ancient Egypt.

It is useful to consider possible artistic representations of filariasis from ancient Egypt but, as always, any scenes suggesting medical conditions must be used with caution. For example, in the Temple of Hatshepsut, Deir el-Bahri, there is a depiction of a trading expedition initiated by the female pharaoh Hatshepsut during the Eighteenth Dynasty. The trading mission, seeking incense and other commodities important to the ancient Egyptians, is shown in a series of reliefs along the temple's inner walls. In one scene we see the Queen of Punt, her husband, and two children receiving an Egyptian emissary prior to starting trading negotiations. This depiction of the Queen of Punt (see the excellent drawing by Richard Parkinson in Filer 1995: fig. 13) has long been the subject of intense debate as to whether or not she had a pathological condition. The queen is depicted with pronounced lordosis (an abnormal inward curvature of the lower spine) and excessive rolls of fat hanging from her arms and legs, although her wrists and ankles are comparatively slim. In the 1960s it was suggested that this depiction of the Queen indicated that she was an achondroplastic dwarf (Kamal n.d.: 15), however, since then a number of possible pathological conditions have been proposed for the Queen's condition (summarized by Filer 1995: 30–1).

Since this (1995) summary of pathologies suggested for the Queen's apparent "deformity," another condition has been added to the list, namely that the Queen may have contracted filariasis, as stated above, a condition spread by blood-sucking mosquitoes (of the genus Culex) wherein larvae of certain worms (transmitted through biting) settle in the lymph nodes, and as stated above, the infection results in (often intense) swelling of certain parts of the body; additionally, there is fever and the sufferer's skin becomes coarse, thick, and fissured, almost elephantine-like. As the worms causing elephantiasis are usually contracted during childhood but do not manifest symptoms for many years (Dobson 2007: 236), this would suggest that the Queen of Punt, if she had contracted elephantiasis, had suffered from the condition for many years. However, as stated above, caution must be adhered to when inferring medical conditions from ancient Egyptian art.

Another ancient Egyptian work of art which has been suggested as representing elephantiasis is a statue of King Mentuhotep (Montuhotep) II, of the Eleventh Dynasty (Dobson 2007: 230); Dobson cites the statue's swollen limbs as suggestive of the condition. This seated sandstone figure of Mentuhotep II (*c.* 2030 BCE) was found accidentally in 1900 when Howard Carter (later to discover the tomb of Tutankhamun) was riding across the forecourt of the king's mortuary temple (at Deir el-Bahri) and the ground beneath his horse's feet caved in revealing the entrance to a chamber below the temple and the subsequent finding of this statue of the king. At a height of 138 cm the statue is in excellent condition and depicts the king sitting with his arms folded. The lower legs and feet, easily seen because the king is wearing a knee-length sed festival garment, certainly do look swollen, over-sized, and disproportionate in comparison to the arms and torso (Russman 1990: 49; Arnold 1999: 104). This figure was once part of a series of statues for this temple and it would have been instructive, had it been possible, to compare this statue with the bodily dimensions of the other statues in the set. However, we can usefully compare this statue with a depiction of the same king on a painted relief fragment now in the British Museum (EA 1397) and originating from the same temple. Here the partially preserved legs (the feet are missing) of Montuhotep II look normally proportioned and not overly thickened. Again, caution must be sounded with regard to interpreting possible pathological conditions from works of art.

As often seen in ancient Egyptian art, bodily proportions can be dictated by the prevailing art style of a period, often reflecting the political status quo of the time. Smith (1998: 85) described the body of the statue cited by Dobson as being treated in a "rude massive fashion" indicating its proportions were part of a particular style of art. Robins (2008: 90) points out that Montuhotep, successful in reuniting his country under a strong government after a period of instability and chaos, seemingly deliberately ordered artwork in the robust style of the earlier Fifth and early Sixth Dynasties in order to recollect former times when Egypt had been strongly and centrally organized. Importantly, she states "male figures exhibit musculature" and thus, it might well be that the "thickened" or muscled legs of the statue cited by Dobson (2007: 230) are really an artistic representation of the king's political power and success and, therefore, not an expression of disease.

In 1828, an address was made by William Osburn to the Leeds Literary and Philosophical Society (England) on the subject of the unwrapping and dissection of a mummy then named Natsef-Amun (also known as the Leeds mummy). Natsef-Amun lived during the Twentieth Dynasty and was a priest during the time of the pharaoh Ramesses XI (*c.* 1100 BCE). The mummy was examined again and re-appraised in 1989 by the Manchester Mummy Team, who decided to retain the original 1828 published reading of the priest's name of Natsef-Amun; the name is now more properly read as Nesamun. During their examination of Natsef-Amun/Nesamun the presence of filarial worms was discovered (Tapp and Wildsmith 1992; Sandison and Tapp 1998: 40–1) in the soft tissues from the mummy's groin area (see Sandison and Tapp 1998: fig. 2.3 for an illustration of this). Might it be suggested that Natsef-Amun/Nesamun had suffered a gross swelling of the scrotum as often seen in filariasis/elephantiasis? Wilkins, as stated earlier, shows modern examples of the appalling genital deformities that this priest may have suffered from in life (1994: fig. 175 left and fig.176 left).

Whatever the situation for filariasis in ancient Egypt it is clear the extreme form of the disease—elephantiasis—if it occurred, would have prevented the sufferer from leading a normal life. This would have been especially so when we remember the importance of contributing manual work to this largely farming society: this was not the case for Natsef-Amun/Nesamun being a priest. However, as pointed out by Dobson (2007: 238), "the deformities associated with the disease lead to social stigma"—was this the case for our middle-aged priest?

In this section on mosquitoes, we have seen some of the devastating effects these insects inflict on humans nowadays and probably did so in ancient Egypt. Here we have considered only two of the disease conditions transmitted by mosquitoes—malaria and filariasis (elephantiasis)—which could have affected ancient Egyptian people specifically and could have had an impact on society generally. We can only hope that the cribra orbitalia and porotic hyperostosis often observed in Egyptian human remains did offer some protection, in ancient times, from deadly mosquitoes which, according to Jones (2012: 10), "are still winning the war" against humans.

As a final comment on the "deadly power" of mosquitoes—it is ironic that Lord Carnarvon, the sponsor of Howard Carter's search for Tutankhamun, reputedly died from a septic mosquito bite, thus adding fuel to the superstition (started in 1923 by the romantic novelist Marie Corelli according to Barker 1972: 32, 33) called the "Curse of Tutankhamun." There again, here the mosquito may be innocent (?) but it certainly has a lot to answer for in terms of general population destruction!

In conclusion, here we have considered just three of the many insects encountered by the ancient Egyptians and, as you can see, there is potentially a considerable amount of information to discuss. There can be no doubt that bees, flies, and mosquitoes affected the lives, particularly the health status, of the populace but in dramatically different ways. As research continues, our knowledge of the part insects have played in both ancient Egyptian (and World) history will expand—insects have so much more to tell us!

CHAPTER THREE

Insects and Food

Evidence from the Archaeological Record

EVA PANAGIOTAKOPULU

The history of insects and food goes back far beyond organized settlements and the first urban centers of the ancient world. Beetles, flies, and other pests would have been accidentally introduced with wild plants and seasonal food resources into sites occupied by hunter gatherers. The association of insects with man-made environments since these initial chance introductions led to their establishment and continuous exploitation of human environments which included storage of any surplus and a variety of artificial suitable alternative habitats. In essence, synanthropy, in other words the dependence on man-made habitats by insects, was simply chance exploitation of alternative resources and subsequent specialization on these. From introducing the first field pests, for example bruchids, together with pulses stored as a seasonal resource, and also sharing insects with rodents and birds in caves to a fauna dependent on the storage of food, this is a story about food, diversification, and complexity. This process is linked with the origins of pests and their first records and the spread of agriculture from the Fertile Crescent to the Mediterranean. Although data are thin partly as a result of limited relevant research from excavations, permits, and lack of preservation from semi-arid sites, there are still interesting results which provide material for an overview based on key sites from pharaonic Egypt, the Aegean, and the Roman world.

STORAGE AND INDOORS ENVIRONMENTS IN PHARAONIC EGYPT

Some of the first studies in archaeoentomology were associated with Egyptian burials, mummies, and tomb offerings. In addition to the early work by Hope (1842), who looked at the contents of the stomach of a mummified Ibis from either Saqqarah or Thebes (Figure 3.1) and produced innovative results on the feeding habits of the bird, other work from mummies in effect produced the first results about insects in stored materials, in this case human bodies. Pasalacqua, an Egyptologist who worked on mummies, was under the impression that the embalming extended to the insects associated with the bodies (Pettigrew 1834), probably as a result of their optimal preservation (see Levinson and Levinson 1996, 2001).

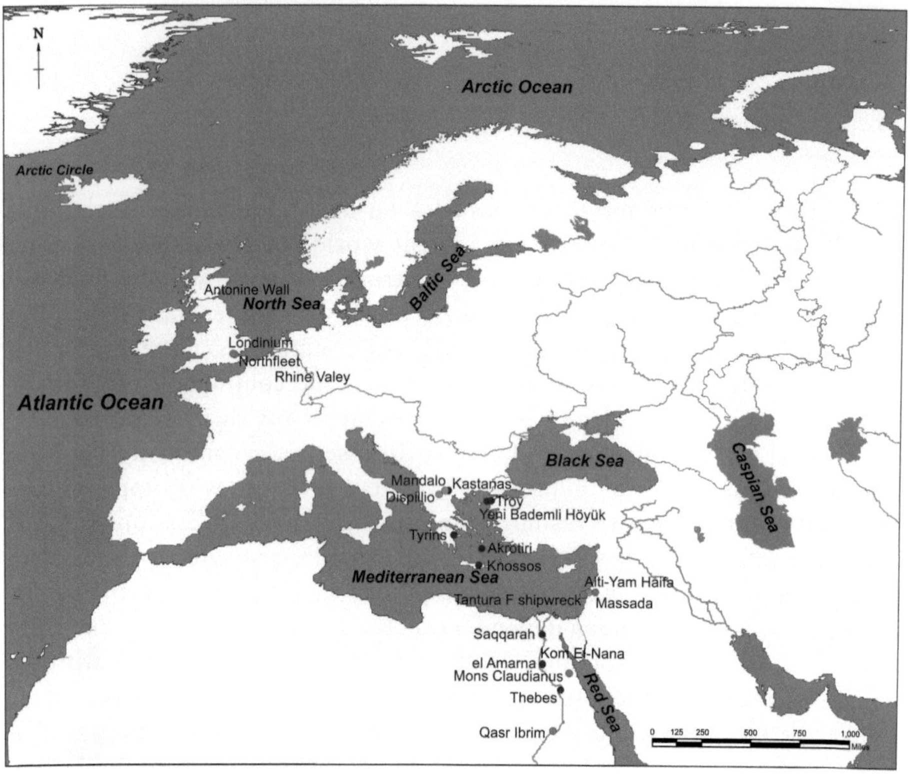

FIGURE 3.1 Location map with relevant sites mentioned in the text. Neolithic: Alti-Yam Haifa, Dispillio, Mandalo; Bronze Age: Akrotiri, el Amarna, Kastanas, Knossos, Saqqarah, Thebes, Troy, Tyrins, Yeni Bademli Höyük; Roman: Antonine Wall, Londinium, Massada, Mons Claudianus, Qasr Ibrim; Post-Roman: Northfleet, Tanura F Shipwreck. © Eva Panagiotakopulu.

Allaud (1908), Alfieri (1931), and Zacher (1934a, 1934b, 1937) produced various papers with results that point to a fauna already introduced in the ancient households which also took advantage of the opportunities presented in mummified remains.

The spell in the Book of the Dead, "'Begone from me, O Crooked-lips! I am Khum, Lord of Peshnu, who dispatches the words of the gods to Re, and I report affairs to their master'" (Book of going forth by day, 36) indicates that the ancient Egyptians were familiar with the relevant fauna and indeed the mummification process was in part to stop the insects from consuming the body (Faulkner 1994). Protecting and preserving seems to be part of the preoccupations with one of the oldest medical documents, the Ebers papyrus (see Ebbell 1937), dated to around 1600 BCE, but probably based on recipes much older than this, where remedies are given to get rid of insects within domestic contexts "fat of gnw bird for biting flies and fresh balanites oil against gnats." Kkt-animals, presumably grain weevils (*Sitophilus granarius* (L.)) which ate the corn in the barns, were controlled by using the burnt dung of gazelle, diluted in water (Ebers Papyrus XCVIII).

From the Book of the Dead to archaeological records, pharaonic Egypt was a culture concerned with storage and preservation. This was also demonstrated in the lengths to which the Egyptians went to ensure securing their crops in a range of different granaries over different periods, either for individual household or communal purposes (Levinson and Levinson 1985, 1994). The buildings would change over time, but the tradition and the motives remained the same—keeping the pest away and minimizing losses through infestation—although this would not necessarily have the expected results. A long list of examples which include discoveries from mummies, tombs, and households indicate that insects found their way into all commodities.

Alfieri (1931) produced some of the first results on pests of stored products from Egyptian tombs from the offerings found in Tutankhamun's tomb who died *c.* 1325 BCE. *Stegobium paniceum* (L.), the biscuit beetle, currently a cosmopolitan pest on a variety of foodstuff which range from caraway seed (Atty 1983) and almond to all kinds of vegetable materials (Dillon and Dillon 1972) and farinaceous foodstuff (Buck 1958), was recovered from the tomb offerings. In addition, the so-called tobacco beetle, *Lasioderma serricorne* (F.), was found: it occurs in a variety of dried vegetable material, spices, and yeast and is also known from tobacco and leather (Attia and Kamel 1965). The lesser grain borer (*Rhyzopertha dominica* (F.)) in the tomb assemblage, also now has a cosmopolitan distribution and is a serious pest in Egypt in a variety of foodstuffs from grain to condiments and malt (Attia and Kamel 1965), was infesting the offerings. The spider beetle (*Gibbium psylloides* Czen.) was also recovered. It has been recorded feeding on a broad range of materials from food and spices to skins (Belles and Halstead 1985).

This early work was complemented by Solomon's finds of *S. granarius* (Solomon 1965) from a tomb near the Step Pyramid at Saqqarah dating from *c.* 2300 BCE, whilst there are also earlier records from *c.* 2900 BCE and from the Sixth Dynasty (*c.* 2323–2150 BCE) (Chaddick and Leek 1972). These were the first records from pharaonic Egypt of perhaps the most important pest of grain worldwide, introduced to early storage areas probably in the Neolithic, and with the earliest records from the pre-pottery Neolithic site at Atlit Yam in the Levant *c.* 6200 BCE (Kislev et al. 2004). It had reached the Rhine valley in Europe by the Linear Band Keramik (LBK) *c.* 5057 BCE (Schmidt 1998).

Species such as the red-legged ham beetle (*Necrobia rufipes* (Deg.)) were recovered from early studies of Graeco-Roman mummies (Hope 1834, 1836; Pettigrew 1834) and later finds include Ramses II (1279–1213 BCE), the Ptolemaic Champollion mummy (Champollion Figeac 1814; Alluaud 1908), and the Bristol mummy (*c.* 1070 BCE) (Miller 1825). One of the so-called bacon beetles (*Dermestes frischi* Kug.) was also recovered in early work on Graeco-Roman mummies and a range of other Egyptian mummies, including animal ones. Both these species were without doubt shared with assemblages from domestic contexts and they have been recovered from settlement sites (Panagiotakopulu et al. 2001), including Tell el Amarna.

From a Ptolemaic mummy, Manchester Mummy 1770, there is the first record of an egg case, ootheca, of *Blatta orientalis* L., the oriental cockroach, which is one of the most common modern pests, particularly in restaurant kitchens; it is a flightless omnivorous scavenger found in a wide variety of different contexts (Marshall and Haes 1988).

Significant information has come to light from a variety of domestic contexts excavated on pharaonic and later sites in Egypt as a result of optimal preservation by desiccation. Many of the insects recovered are associated with human and animal provender and other remains in and around households. From granaries to floor deposits and middens or eroding mudbrick, insect pests provided interesting details about everyday life and trade in commodities.

At Tell el Amarna the broad range of pests on foodstuffs (Figure 3.2) from the site provides an indication of both the issues with infestation in food and the extensive network and contacts during the New Kingdom. One of the most frequent pests was the endemic lesser mealworm (*Alphitobius diaperinus* (Panz.)), a pest that is omnivorous and can be found on commodities which range from grain to leather, and flour to ground nuts (Brendell 1975). There is abundant evidence for infestations of *Coccotrypes dactyliperda* (F.), the date stone beetle, and large numbers of infested date stones are frequently recovered (Figure 3.3). In addition, the long list of pests from the settlement includes *S. granarius*, *R. dominica*, *G. psylloides*, *S. paniceum*, and *D. frischii.* The flour beetle (*Tribolium castaneum* (Hbst.)), a secondary pest in bran and flour, and found in a variety of additional grain products (Andres 1931), also

FIGURE 3.2 *Alphitobius diaperinus* (Panz.) from the Workmen's village, Amarna, Egypt. © Eva Panagiotakopulu.

FIGURE 3.3 Infestation of date palms from Amarna, Egypt, with visible holes from *Coccotrypes dactyliperda* (F.). © Eva Panagiotakopulu.

occurs, along with other secondary pests such as the small-eyed flour beetle (*Palorus ratzeburgi* (Wiss.)), currently frequently found in flour and mouldy grain residues attacked by other pests, primarily the granary weevil (Brendell 1975). The flat grain beetle (*Cryptolestes turcicus* Grouv.) has similar preferences for cereal products, but is also found in dried fruits (Halstead 1993) and the long-headed flour beetle (*Latheticus oryzae* Water.), which also infests cereal products (Fogliazza and Pagani 1993), form part of the list of pests from the site. *Trogoderma granarium* Everts, the khapra beetle, an important warm temperate to tropical pest of stored products, in Egypt appears to be found primarily in grain and flour, but also recovered from various vegetable products (Khalifa and Badawy 1960). The seed beetles (*Bruchus* sp. and *Bruchidius* sp.) also found their way into living areas, probably together with animal fodder and stored pulses. Although there is limited research from sites earlier than the New Kingdom, first results from Middle Kingdom Elephantine reveal similar assemblages, indicating that introduction of several of these pests with origins beyond the Nile valley took place much earlier, perhaps at the beginnings of the Neolithic and their establishment and further spread took place with the initiation of the pharaonic state. From *S. granarius*, with probable origins in the Fertile Crescent, to *R. dominica*, perhaps of Indian origin, *T. castaneum*, perhaps from the Levant, to Egyptian endemic species, such as *A. diaperinus* and *G. psylloides*, the interactions and trade of goods responsible for the initial introduction of the fauna found in these domestic areas must have had a significant impact on the development of complex societies in the Nile valley.

Given the long list of insect pests, the infestation levels in granaries, which could lead to the total loss of crop in storage, would have significant effects particularly during poor harvest years and could even result in famine and facilitate outbreaks of disease in weakened communities. The movement of food commodities along the Nile and beyond is another significant point of the overall story. Limited research from sites in Iran by Costantini and colleagues (Costantini et al. 1977) indicate a range of pests which include *S. granarius* and *G. psylloides* from 2600 BCE at Shahr-i Sokhta in Baluchistan showing the range of these networks which also extended northwards to the Aegean and beyond.

THE AEGEAN AND GREECE

Research from the Aegean and northern Greece has produced some of the earliest pest records in Europe, indicating the importance of the broader region for the spread of storage-based farming. The grain weevil (*S. granarius*) appears at Dispilio in Northern Greece *c.* 5700 BCE and is the earliest European record of this flightless pest. Insufficient research has been done in the intervening regions, but it provides information about the possible itinerary of the invasive species perhaps from the Levantine coast over to the Aegean and northern Greece before finding its way, probably with the initial seed corn, to the Rhine valley during the LBK (Panagiotakopulu and Buckland 2018). In addition, the saw-toothed grain beetle (*Oryzaephilus surinamensis* (L.)), a pest of cereals and cereal products, although it infests a wide range of commodities (Halstead 1993), was recovered from charred grain from a late Neolithic site at Mandalon (4450–4340 *cal* BCE) in northern Greece (Valamoti and Buckland 1995), perhaps spreading in the opposite direction from a preferred natural habitat in *Urwald* (Panagiotakopulu and Buckland 2017). Early field pests in the Aegean include bruchids, although in most cases the identifications have been only to genus level. A range of sites from Asia Minor to the Peloponnese provided examples of bruchids. These data include records from early Bronze Age (3000–2200 BCE) and middle Bronze Age Troy (2200–1700 BCE) (*Bruchus* sp.); an early Bronze Age site at Yenibademli Höyük, Turkey, *c.* third millennium BCE; late Bronze Age Kastanas *c.* 1600–1200 BCE (Bruchidae); and Mycenaean Tiryns *c.* 1200 BCE (Bruchidae). These records contribute to the discussion of the spread of cultivation of pulses which may predate cereal cultivation (Panagiotakopulu et al. 2013).

The only site studied in detail in the Aegean is Akrotiri, a late Bronze Age site on Thera (Santorini) destroyed during a Plinian volcanic eruption around BCE (Doumas 1983). It provides interesting data about pests and contributes to the understanding of the biogeography of the region, as most of the finds from the site are pests. From storage pithoi with *Lathyrus clymenum* L. there was evidence of bruchid infestation. The species is identified as *Bruchus rufipes* Hbst. (Figure 3.4), a species which feeds on *Vicia* spp. (Delobel and Delobel 2006), but also found on *Lathyrus* spp. (Anton et al. 1997). This bruchid is univoltine and does not develop in dry stored pulses but may be introduced to storage in infested field crops. Akrotiri also provides data about levels of infestations in crops of fava beans, which is calculated to have been around 8–9 percent. It was evident that in at least one case the crop was sieved and separated to prime and tail grain and stored in different pithoi, and in different parts of Room 5, where storage was taking place in the West House (Sarpaki 1992). In the pithos with the prime grain, the presence of *Coriandrum sativum* L. seeds indicated the use of insecticides on site. Other plants indicated were *Thymelea hirsuta* L. and *Echium* sp. which could repel insects but not exterminate them. Although other substances, such as wood, ash, or even fishbones could have been used for the same purpose (Panagiotakopulu et al. 2001), it is virtually impossible to confirm this in most excavated contexts. As a result of the unique preservation of the site and taking into account the life cycle of the beetle, which takes fifty to sixty days from egg to adult (Genduso 1960), in relation to cultivation of the crop (*L. clymenum*) and the bruchids infesting the crops at Akrotiri, a season

FIGURE 3.4 Fava beans (*Lathyrus clymenum* L.), from the West House, Akrotiri, Santorini, Greece infested with bruchids (*Bruchus rufipes* Hbst.).

for the volcanic eruption, during the summer, perhaps June or July, has been proposed (Panagiotakopulu 2000; Panagiotakopulu et al. 2013).

Other main pests at Akrotiri include *S. granarius*, *L. serricorne*, *O. surinamensis*, and *T. castaneum*, all introduced with trade and established at Akrotiri in crop storage. These show that pithoi were used for diverse materials and perhaps not effectively cleaned between uses. In addition, the range of the fauna indicates that the Aegean islands were part of a larger interconnected system. Two blind flightless beetles, *Troglorhynchus anophthalmus* Schmidt and *Amaurorhinus bewickianus* Woll., both improbable pests, were also found from these assemblages. Troglodytic, they probably show that storage in caves had taken place or that stored crops were covered with leaf litter (Hoffmann 1954).

Although the grain weevil (S. *granarius*) was recovered from Late Minoan II Knossos (Jones 1983), there has been little further research in Greece, partly because preservation tends to be poor. However, there is extensive mention of infestation in Classical sources (Beavis 1988) and evidence for the use of insecticides in order to deter pests from devouring stored foodstuffs (Panagiotakopulu et al. 1995). Some of these include oils, such as olive oil, mentioned by Aristotle HA VIII 27, Hipp. Cant. LXXI 14, often mixed with other substances, particular earths, as for example clay, proposed by Theophrastus (DCP VI). During the Hellenistic period the general view was that insects disliked strong smells and tastes and this opinion was shared by Roman agricultural writers who also documented the severe problems from pests in the Roman storerooms.

THE ROMAN ARMY

It is unfortunate that suitable sites to document the spread of pests in the Hellenic and Hellenistic periods have not been examined—it would be interesting, for example, to know what was in and what was hitchhiked in Alexander's baggage train—the next period with significant data is the Roman. Although trade in perishable commodities had led to the spread of insect pests from the Fertile Crescent and Near East to northern Europe at least temporarily in the Neolithic, it was not until the Roman period that they were spread to the more remote parts of the Empire, including the British Isles. The need to feed the Roman army was a driver for the extensive use of large granaries and concomitant large-scale insect pest infestations.

Agricultural writers during the Roman period also discussed problems with infestations and ways of repelling pests. From using olive oil (Aelian IV:18) to storing in oil jars which were covered with ashes to placing crops on top of vinegar jars or drying crops in the sun and then using vinegar among other substances (Columella, RR II 10 16; Pliny, HN XVIII lxxiii 308). *Amurca*, which consisted of olive oil lees mixed with soda, seems to be the most frequently mentioned insect repellent in the Roman literature (Pliny HN XVIII lxxiii 305; Cato, DA XCV, XCVI, XCII). From plastered casks, which had contained salted fish to ashes, wormwood, charcoal, and cow dung, the list of natural remedies is long (Pliny HN XIX lviii 178, Virgil Georg. I178, Palladius VIII, 3.2). It also involved "chalk" from specific locations (Chalcis, Caria) whilst earths, probably high in clay, were also recommended (Pliny, HN XVIII 305; Varro, RR I LVII), and these may have included substances such as diatomites which would have some impact on infestation levels. Storage in caves (Varro, RR I LVII), in pits (Pliny, NH XVIII, 306), covering their base with straw, and storage of crops in the ear (Pliny, HN XVIII lxxiii 306) were also used to control infestation. Although the list of repellents is long, they tend to limit as opposed to eliminate pests and often enough infestations may be so extensive that little can be done apart from burning the infested crop in order to limit the extent of the damage to future storage. In spite of the difficulties of identifying these instances in the archaeological record, the burnt grain from the east Yorkshire Roman fort at Malton provides an interesting example (Buckland 1982).

In terms of pests and their spread during the Roman period, from the granaries of Egypt to the silos of Rome and from Rome to the most remote settlements, the route to imperialism included a brand new ecological footprint and introduced species, both purposefully and by chance, played a significant role.

Research in Roman Egypt is limited to a handful of sites studied for insects but they present good evidence about changes in diversity and new pests introduced from the start of the Roman period onwards. One of the main changes in terms of insect faunas associated with food is the sheer numbers of individuals from the relevant contexts, perhaps reflecting the extent of bulk storage, and evidence for processed food, such as flour and biscuits from relevant assemblages. This includes remote desert sites like Mons Claudianus (Panagiotakopulu and van der Veen 1997), and Qasr Ibrim on the Nile in Upper Egypt (Panagiotakopulu forthcoming).

In terms of introductions of new pests, the extent of accidental guests is exemplified from finds from sites in the Egyptian desert. From the Roman quarry of Mons Claudianus, in addition to granary weevils and the red flour beetle (*T. castaneum*) hitching a hike with flour and biscuits for the Roman soldiers, there was evidence for the introduction of another exotic pest, *Aglenus brunneus* Gyll. This is currently cosmopolitan, known from a variety of materials including mouldy hay, straw, and stored grain (Hinton 1945; Koch 1989). Since the beetle is blind and flightless, its presence in Roman Egypt would be associated with its on accidental transport with food materials, probably infested and mouldy barley brought over for the army's beer.

There is a similar story from the fort of Masada in Galilee, which was destroyed by the Romans around 73 CE (Kislev and Simchoni 2007). The species list includes the granary weevil, the flat grain beetle (*Cryptolestes ferrugineus* (Steph.)) and the red flour beetle. In addition, the cigarette beetle, lesser grain borer, and the date stone beetle were all recovered from Masada, indicating similarities with other assemblages recovered during the same period.

The three most significant storage pest introductions to northern Europe, which the Romans are responsible for, are the granary weevil, *S. granarius*, the saw-toothed grain beetle, *O. surinamensis*, and the flat grain beetle, *C. ferrugineus*, the primary triumvirate of grain pests during the Roman period. *C. ferrugineus* has its earliest records during the Roman Empire and probably this is the point that it moved into the storerooms from its primary niche, under bark (Halstead 1993), in order to exploit grain and food residues. The spread of all three pests was significant, with records from the Levant (Kislev and Simchoni 2007) over to the most northerly frontier in the British Isles, the Antonine Wall (Smith 2004; Locke 2016). From the British records of *S. granarius*, there are some interesting records about the spread of the pest. The granary weevil tends to be absent from the rural sites and frequent in towns and forts, reinforcing its strong association with bulk storage of grain. In addition, the movement of the pest in Britain mirrors the itinerary of the

Roman army, from Londinium to the Antonine wall, giving clear evidence for its spread (Panagiotakopulu and Buckland 2017). *O. surinamensis* is found together with *S. granarius* at most sites. The red flour beetle (*T. castaneum*) is another species strongly associated with the food of the Roman army. Processed food, flour, and biscuits are likely to have been part of the supplies distributed to forts and this probably facilitated the spread of the flour beetle to northern Europe during the Roman period. Their distribution diminishes during the post Roman period. An additional parameter could have been favorable climatic conditions and warmer temperatures (cf. McMichael 2011) which could have created optimal conditions for the insect introductions. These would have aided the establishment of pests in northern parts of the Empire, although the microclimate within the areas where the foodstuffs were stored would have been more important than macroclimate.

POST-ROMAN PERIOD TO THE MODERN TIMES

Although evidence after the Roman period regarding insect pests on food comes primarily from northern Europe, there are some data from Egypt. An interesting introduction is that of the saw-toothed merchant grain beetle (*Oryzaephilus mercator* (Fauvel)) from sixth-century early Christian Qasr Ibrim, in Lower Nubia. This is the first record of the species which is now tropicocosmopolitan and prefers nuts and oil seeds. At Qasr Ibrim it was found in numbers in association with radish seeds, which point to this pest being established during the Roman period, when radish was used by the Roman army for oil. The grain weevil, flour beetle, and *A. diaperinus* were also recovered from the fifth–ninth century CE Kom el Nana. In addition, the khapra beetle (*T. granarium*) was also recovered from both sites providing early fossil records for this now cosmopolitan pest of stored cereals; its probable origins lie in southeast Asia or India (Peacock 1993). The pest was recovered in high numbers from Roman and post-Roman Qasr Ibrim (Figure 3.5), perhaps indicating changes in either trade networks or storage regimes. Apart from these two sites, there is evidence for the granary weevil at Tantura, in an early Islamic shipwreck off the Levantine coast (Barkai et al. 2013).

An overview of the fate of post-Roman introductions shows that, in most cases, they disappear from northern Europe. For example, the fossil record of the pest fauna of the British Isles, with the exception of Northfleet in Kent (Smith 2011), dwindles, probable as a direct example of lack of shipments of grain and other changes after the fall of the Roman Empire during the fifth century CE. Although evidence is limited, there must have been places where the grain fauna persisted, perhaps on sites where grain was milled. During the medieval period there was a renewed increase in sites with faunas

FIGURE 3.5 *Trogoderma granarium* Everts from Qasr Ibrim. © Eva Panagiotakopulu.

indicating pests in food, even compared to the Roman period. Again in this case ameliorating climate might have played a role in creating optimal conditions for the spread of the grain fauna. The increase in number of sites applies to the main grain pests, *Sitophilus granarius* and *Oryzaephilus surinamensis*, but not to *Tribolium castaneum*. Although research during this period comes primarily from the British Isles, it is quite evident that processed grain stored foods were limited, in comparison to supplies during the Roman period. The medieval period is followed by another period where occurrences of pests became much lower, although few sites have been examined. Perhaps the Little Ice Age and climate cooling (McMichael 2011) and also the effects of Plague may have been reasons for this.

Perhaps a very significant change during the medieval period was the transition from ships which were beached and emptied of ballast and dunnage to vessels more permanently afloat; these would only be fully offloaded occasionally, allowing the grain faunas and other pests, such as the house fly (*Musca domestica* L.) to travel on board for longer periods, and eventually allowing the development of more cosmopolitan assemblages of synanthropic insects.

CONCLUSIONS

The results from the fossil insect record provide an overview of the spread of grain pests which nowadays have become cosmopolitan. From pharaonic Egypt to remote parts of the Roman Empire to the medieval period and beyond, accidental introductions of insects accompanied trade of crops and food materials often associated with the movement of immigrants and armies. The footsteps of primary and secondary pests can be reconstructed over time through their fossil records. These introduced species have shaped a new world with homogenized grain faunas, where food losses and their consequences become one of the main drivers for the intensification of farming and where significant infestations might lead to famine and disease.

CHAPTER FOUR

Insect Products

GENE KRITSKY

The use of insect products began long before the start of antiquity, and the oldest example involves insects that we still utilize, honey bees. Pottery shards, found in Anatolia that date back nearly nine thousand years, contained the chemical residue of beeswax (Roffet-Salque 2015). It is assumed that honeycomb was placed in these containers presumably to use the honey as food, but there is no evidence of how the wax was used. However, this discovery does not establish that beekeeping was being practiced at the time, as honeycomb can also be obtained by honey hunting. What it does establish is that humans profited from the products of insects long before recorded history.

Honey hunting is the opportunistic stealing of honeycomb, which is very different from beekeeping. Beekeeping is the intentional provision of an artificial cavity within which bees will build comb, deposit nectar to process into honey, and rear the young (Kritsky 2010). Beehives are made by humans for the bees to use, their creation was necessary for us to increase our access to honey bee products. The oldest definitive evidence of beekeeping is a Fifth Dynasty carved bas-relief that dates to 2450 BCE from Egypt. Unfortunately, the relief shows an advanced beekeeping operation already in place. It does not provide evidence when beekeeping developed in Egypt. It is likely that it was being practiced as early as 3000 BCE, as evidenced by crude honey bee hieroglyphs carved during Pre-dynastic times (Kritsky 2015).

The first beehives may have been the result of a swarm of bees taking refuge in a large empty container. Honey hunters may have noticed the bees in residence in the once-empty container and decided to encourage more bees to seek safe harbor by placing additional containers in the vicinity of other bees, completing the transition from honey hunting to beekeeping.

Early beehives used in antiquity were quite different from the box hives we see today. The hives illustrated on the Ancient Egyptian relief were horizontal hives made of mud and plants that were rolled into a long tube which was dried and then stacked upon each other forming a walled apiary. A painted beekeeping scene from Egypt's New Kingdom (1500 BCE) illustrates how the round honeycombs from these tube hives were removed, crushed into a mound of honey and wax, and placed into a tall vessel likely to separate the honey from the wax. The resulting honey was then stored in diamond-shaped containers made by placing a bowl upside down upon another bowl and sealing the rims with bees wax to keep the honey from spilling out (Kritsky 2015).

Beekeeping practices also evolved during antiquity. The earliest representation from Egypt's Fifth Dynasty shows a beekeeper working the hives with some kind of container near his face (Figure 4.1, far left). Some have suggested this was a smoker and others suggest it might have been involved in calling the bees. The hieroglyphs above the relief read "to emit a breath or to make a little sound." If the former interpretation is accurate, then the scene could be a beekeeper blowing into a smoker except the container is unlike the incense bowl that is used in the Eighteenth Dynasty painting at Rekhmire (Figure 4.2). If the inscription reads to make a sound, then it could represent the beekeeper calling the queen. When new queens emerge from their cells, they make a sound by vibrating their wing muscles without moving their wings producing an audible "zeep" sound. Other developing queens still in their cells will respond with a different sound. Traditional Egyptian beekeepers would mimic the call of the new queen and listen for developing queens. If they should hear the response of a queen still in her chamber the beekeepers would open the hive and remove the comb with the sealed queen and place it into an empty hive along with some workers to start a new colony. Today this is referred to as "artificial swarming" (Kritsky 2015).

FIGURE 4.1 The beekeeping relief from the Solar Temple of Newoserre Any. Reconstruction by Gene Kritsky.

FIGURE 4.2 A reconstruction of the beekeeping painting in the Tomb of Rekhmire, Eighteenth Dynasty. At the far right is a beekeeper holding a dish of incense to quiet the bees. Reconstruction by Gene Kritsky.

The next scene in the Egyptian Fifth Dynasty relief (Figure 4.1) shows three men decanting honey into larger vessels. One of the containers has a spout that was placed along the side of the container. This may have been a method to separate honey from wax. When honeycomb is crushed and placed in a container, the wax will float to the surface and the heavier honey will be below the wax. The spout on the side permits the pouring of the honey into the container while leaving the wax, which is floating at the top of the container, in the spouted vessel (Kritsky 2015).

The third scene is badly damaged showing only two beekeepers presumably with a vessel between them. The inscription reads "to press." This may be illustrating further separation of the remaining bits of honey that remained with the wax. The fourth scene, which is complete with some of the original color, shows a beekeeper sealing honey in bound vessels for storage (Kritsky 2015).

The painted scene from the tomb of Rekhmire adds a few more details to Egyptians' process of honey extraction. In particular, beekeepers are using smoke from an incense burner as an offering to the bees which would have quieted the bees. The round honeycombs taken from the hives are crushed into a mass of wax and honey which is placed into a container, and is subsequently placed into larger vessels (Figure 4.2, from right to left).

Honey and beeswax were so important to the Ancient Egyptians that beekeepers were organized into a hierarchy that monitored honey production and answered to the Pharaoh via his vizier. These positions included beekeepers, chief beekeepers, overseers of the beekeepers, and so on. After Ancient Egypt fell under the control of Rome, beekeepers were organized into a guild (Kritsky 2015).

Beekeeping was also highly organized in other parts of the Fertile Crescent; although the evidence is indirect. The Hittite Kingdom, which occupied much of Anatolia (Modern Turkey), had laws and penalties to discourage the theft of beehives. Cuneiform tablets dating to 1500 BCE document that the theft of a swarm of bees would have the thief pay the hive owner five shekels of silver. This fine shows that bees were valuable as the penalty for stealing a goat was also five shekels. If the bee rustler heisted two to three beehives, the penalty was six shekels. Even the theft of an empty hive would result in the fine of three silver shekels. The Hittite laws also record that the theft of a container of honey was one shekel, the same value as a swarm or a quantity of butter (Crane 1999; Kritsky 2017). There is no description of the beehives used, but the law does indicate that an empty hive could be stolen.

In 2005, a team under the direction of Dr. Amihai Mazar from the Institute of Archaeology at Hebrew University of Jerusalem in Israel discovered eight beehives at Tel Rehov. An additional seventeen beehives were excavated in 2007 (Mazar et al. 2007). These were the most definitive ancient beehives to be discovered. Carbon dates place the hives between 970 to 840 BCE, during the time of Solomon and the subsequent division of the Kingdom following Solomon's death (Mazar et al. 2008). Tel Rehov is not mentioned in the Bible, but it is mentioned in several Egyptian texts that date to the New Kingdom. These texts state that Tel Rehov remained loyal to the Pharaoh Seti I while neighboring towns revolted. Visitors to Karnak Temple in Luxor, Egypt will see Tel Rehov mentioned in the list of towns that Pharaoh Shoshenq I conquered during his campaign into region (Mazar 2007).

The Tel Rehov hives somewhat resemble the beehives illustrated in the New Kingdom tomb of Rekhmire (see Figure 4.2), being large cylindrical hives with three hives stacked upon each other. The major difference is the rear of the Rekhmire hives are bluntly rounded, whereas the hives at Tel Rehov were capped with at mud lid that fit the opening (Mazar et al. 2008). The verification that these structures were indeed beehives comes from the chemical residues of beeswax and parts of honey bees that were found within the hives. These bee remains included a wing, which retained characteristics of the subspecies (*Apis mellifera anatoliaca*), which currently is found in cooler parts of Turkey. The subspecies (*Apis mellifera syriaca*) is the honey bee that occurs in the region of Tel Rehov today (Bloch et al. 2010). The discovery of a different bee subspecies suggests either *syriaca* evolved in the last three thousand years or that honey bees were imported into the region. The suggestion of importation is not new to the ancient world. A century after the hives were in use at Tel Rehov, the Assyrian Governor of Suhu, Shamash-res-usur, imported bees from cooler mountainous of Habha (Crane 1999). An Assyrian relief dating to 780 BCE reads:

> I am Shamash-res-usur, Governor of Suhu and Mari. Bees which gather honey, which no one among my forefathers had seen nor brought down from the land of Suhu, I brought down from the mountains of Habhu and established in the town Gabbari-ibni. They collect honey and wax. I understand how to

do the melting out of the honey and wax, and the gardeners also understand it. Any later person who appears, let him ask the old men of the country whether it is true that Shamash-res-usur the Governor of Suhu introduced bees.

(Saggs 1984)

The connection between bees and plants may have been understood by Egypt's Twenty-Sixth Dynasty. At both the tombs of Pabasa and Ankkhor are beekeeping reliefs that include sealing honey in containers, offerings to the bees, and a relief of a beekeeper beckoning bees to trees (Kritsky 2015). This connection between bees and plants was confirmed from a 300 BCE poem that connected honey bees to the sun god Re. The poem reads:

The god Re wept and the tears
from his eyes fell on the ground
and turned into a bee.
The bee made (his honeycomb)
and busied himself
with the flowers of every plant;
and so wax was made
and also honey
out of the tears of Re.

(Leek 1975)

The oldest beehives discovered in ancient Greece date to around 400 BCE. These hives were horizontal hives like those illustrated in Egyptian temples and tombs and found at Tel Rehov, but were made out of pottery rather than mud. In the storage collection at the Agora Museum in Athens and on display in the Museum at Athens International Airport are examples of such pottery beehives (Figure 4.3). These hives were slightly more than half a meter long with a diameter of approximately 28 cm. They had a large opening at one end and were closed off at the other. The opening was furnished with a lipped rim that served to hold the rope that secured a round cover over the opening. The rope was placed behind the rim of the hive and tied to a stick that was placed in front of the hive's cover. Many of these hives were scored on the interior to provide more surface area for the bees to better secure the honeycomb to the hive. The Greek hives could be extended by placing pottery rings between the cover and hive body, which would provide a method for the beekeeper to control swarming.

The Greek pottery hives were occasionally repurposed as coffins for small children. Excavations have found examples where an infant would be enclosed within a hive or a larger child would be interred in two hives that were oriented

FIGURE 4.3 A Greek pottery hive, 300–100 BCE. Photograph by Gene Kritsky.

with the openings held together at the middle by ropes securing the opening of the two hives together (Jones 1962).

Although the oldest Greek hives are only 2,400 years old, there is firm evidence that the ancient Greeks were working with bees a millennium earlier. A pottery hive smoker was discovered in Sesklo dating to 3500 BCE. A larger similar smoker was found on Crete dating between 1750 and 1450 BCE (Kritsky 2018). These were the first specially made devices to produce smoke to quiet bees. Greek smokers were equipped with handles for the beekeeper to hold or to insert a piece of wood to hold the smoker. Holes on the sides of the pot would provide air for combustion and to direct the smoke out of the pot (Figure 4.4).

Beekeeping continued to flourish during the Roman Era. Horizontal pottery hives dating to 200–300 CE were discovered in the Iberian Peninsula. Unlike the closed-end pottery hives the Greeks were using, these Iberian hives were open at each end and were likely fitted with covers or plugs to give the bees an enclosed space (Bonet Rosado et al. 1997) (Figure 4.5).

Books describing Roman beehives help us appreciate the evolution of beekeeping during this period. The author Varro (116–27 BCE) recorded details of differently shaped hives made from a variety of materials:

> Some build round hives of withies (wicker) for the bees to stay in, others of wood and bark, others of a hollow tree, others build of earthenware, and

FIGURE 4.4 A 3500 BCE pottery smoker from Sesklo. Photograph by Gene Kritsky.

FIGURE 4.5 Roman-Era pottery beehives that were open at each end that would be closed with a pottery cover. Photograph by Gene Kritsky.

> still others fashion them of fennel stalks, building them square, about three feet long and one foot deep, but making them narrower when there are not enough bees to fill them, so that they will not lose heart in a large empty space ... The best hives are those made of bark, and the worst those made of earthenware, because the latter are most severely affected by cold in winter and by heat in summer.
>
> (Crane 1999)

Varro also reported that hives were placed in niches in walls for protection. It is noteworthy that he wrote that pottery hives were less effective than those made of bark, possibly cork, as upright oriented cork hives were common in many localities along the Mediterranean Sea as recently as the last century.

Columella (4–70 CE), in his book *De re Rustica*, went into more detail than other Roman authors providing details on the races of bees, swarms, and bee diseases and pests. His description of the various types of beehives suggest that beekeepers were experimenting with hive materials.

> Beehives are to be made according to the character of the region. If there is an abundance of cork trees, we shall certainly make the most useful hives from their bark, because they will not be cold in winter nor hot in summer. If there are plenty of fennel stalks, good vessels for beehives can be constructed out of them, because they have similar properties of cork. If neither of these materials is available, they can be woven out of withies by means of basketwork. If this material too is not available, they will be made out of wood, either a hollow tree or form wood cut into boards. The earthenware hives have the worst properties, because they are burnt by the summer heats, and frozen by the winter's cold. Two kinds of hives remain to be described, those which are made of dung or built of bricks.
>
> (Crane 1999)

Columella wrote that withies, or willow, using basket work may indicate that the first skeps, or basket hives, were in use by this time. The oldest skep hive was discovered in a peat bog near Wilhelmshaven in Lower Saxony in what is northernmost Germany today. It was made of wicker and has been dated to 200 CE (Kritsky 2017).

We have a much more complete understanding of how honey and wax were harvested during Ancient Rome from the writings of Columella. Two tools were used to harvest the honey. Both were about 18 inches long and were equipped with different functional ends, one possessed a sharp, flat edge at the front to cut comb off from the sides of the hives, and the other a curved scraper at the end to remove any remaining comb. These tools survive to this day as the standard hive tools.

Columella goes into detail about how the combs should be removed, and how to prepare the bee-house where the honey will be extracted to discourage robber bees from getting into the bee-house. He also recommended that the honey extraction occur on the same day the combs were cut. To separate the honey from the comb, he wrote:

> A wickerwork basket or a bag rather loosely woven of fine withies in the shape of an inverted cone, like that through which wine is strained, is hung up in a dark place, and then the honey-combs are heaped in it one by one. But care must be taken that those parts of the waxen cells, which contain either young bees or dirty red matter are separated from them, for they have an ill flavor and corrupt the honey with their juice. Then, when the honey has been strained and has flowed down into the basin put underneath to catch it, it is transferred to earthenware vessels which are left open for a few days until the fresh produce ceases to ferment; and it must be frequently skimmed with a ladle. Next the fragments of the honeycombs, which have remained in the bag, are handled again and the juice squeezed out of them. What flows from them is honey of the second quality and is stored apart by itself by the more careful people, lest any of the honey of the best flavor should deteriorate by having this brought into contact with it.
>
> (Crane 1999)

Columella also described the collection of beeswax:

> The yield of wax, though of little monetary value, must not be overlooked, since its use is necessary for many purposes. The remains of the honeycombs, when they have been well squeezed, after being carefully washed in fresh water, are thrown into a brazen vessel; water is then added to them and they are melted over a fire. When this has been done, the wax is poured out and strained through straw or rushes. It is then boiled over again a second time in the same manner and poured in such molds as one has thought suitable, water having been first added. When the wax has hardened, it is easy to take it out, since the liquid which remains in the bottom does not allow it to stick to the molds.
>
> (Crane 1999)

The primary product of beekeeping is honey, but not all honey is the same and these differences were well discussed in antiquity. There were three notions about where honey came from. Many held that it fell from the air. Another idea considered honey was the "sweat" of oaks, whereas a third group believed honey was gathered by bees from the flowers. Seneca the Younger, a Roman philosopher who lived during the first century of the Common Era, thought correctly that bees changed nectar into honey (Fraser 1931).

Honey is made from plant nectar, a dilute sucrose solution, that is collected by bees. The amount of effort it takes to produce a simple kilogram (2.2 pounds) of honey is staggering. Flowers produce so little nectar that it requires a bee to visit approximately a thousand flowers to gather a nectar load, the amount of nectar a single bee can return to the hive. It takes approximately a hundred thousand nectar loads to eventually produce 1 kg, which would require a hundred million flowers! That would require the bees to collectively fly, on average, about 240,000 miles just to produce 1 kg of honey.

Once the bees return the nectar to the hive, the nectar must be mixed with enzymes and the excess water removed. As the forager bee arrives at the hive, it regurgitates the nectar that is consumed by a hive bee. The hive bee then regurgitates little droplets of nectar in and out of its mouthparts to start the evaporation of the water and to mix it with enzymes to convert the sucrose into fructose and glucose, and the later sugar is converted into gluconic acid by a second enzyme. The enzyme-infused nectar is then placed in a cell in the honeycomb. Meanwhile, other bees are beating their wings, creating airflow, and speeding up the evaporation of the water, thereby thickening the honey.

Finally, the honey is sealed in the comb, where the hive's 95° F temperature cooks the honey to remove any plant toxins. The color of the honey is determined from pigments in the nectar the bees collected, and the "bouquet" is determined from the essential oils also produced by the plants (Berenbaum 2010).

The Ancient Egyptians recognized three grades of honey based on where it was collected and its color. Red honey, for example, was desert honey (Kritsky 2015). The Roman author Pliny (XI.15) writes that the reddish-gold thyme honey was a superior honey to wild honey, which was also called heather honey. Columella (Crane 1999) wrote, "thyme yields honey with the best flavor; the next best are Greek savory, wild thyme, and marjoram. In the third class, but still of high quality are rosemary and our Italian savory .... Next the flowers of the tamarisk and the jujube-tree and other kinds of fodder which I suggested have only mediocre flavor. Honey which is considered the poorest quality is the woodland honey."

Honey was an important food product during antiquity. During Egypt's Nineteenth Dynasty, it was reported that honey was provided as a basic ration for the King's Messenger and Standard Bearers (Kritsky 2015). The Greek Pythagoras, who lived *c.* 570–495 BCE, claimed that consuming honey was necessary for a long life. Homer's *Odyssey* wrote that Circe lured Ulysses' men to her house and "she set them upon benches and seats and mixed them a mess with cheese, honey, meal and Pramnian" (Crane 1999).

Honey was also an ingredient in ancient recipes. In the tomb of Rekhmire, next to the painted scene showing the harvest of honey, is a painting of honey being used to make honey and sesame cakes (Kritsky 2015). Honey cakes were also common in ancient Greece. The ancient cookbook pseudoepigraphically attributed to Apicius (2009), a Roman gourmet who lived during the reign of Tiberius in the first century of the Common Era, mentioned honey 273 times. Honey was mixed with spiced wine or with wine made with rose petals as a refreshing drink; it was added to improve the flavor of soup; it was used to sweeten cakes; and to make sauces for truffles, duck, and fish.

Honey's antibacterial properties were unknown during antiquity, but it was recognized in antiquity that honey would stop spoilage. We now know that honey has a low moisture content and low pH that results in honey being antibacterial. Apicius recommended honey to preserve meat. Moreover, various fruits, such as cherries, plums, apples, pears, and figs, could be stored in honey.

If honey could preserve foodstuffs, it could also preserve a corpse from corruption. The Babylonians used honey for burials before 1100 BCE. Herodotus recorded that Democritus, a philosopher who lived in Greece in the fourth century BCE, was buried in honey. There are several accounts that Alexander the Great, who died in 323 BCE, was also preserved in honey. The oldest of the Alexander accounts dates back to first century of the Common Era, and they continued for a millennium. One of the later stories claimed that Alexander himself ordered that a lead coffin filled with honey, myrrh, and the oil extracted from roses be prepared to preserve his body (Crane 1999).

Honey was an important pharmaceutical throughout the ancient world. In Egypt, over nine hundred medical treatments are described in the several medical papyri and honey figures as an ingredient in about five hundred of them (Kritsky 2015). It was used to treat cuts, burns, digestive problems, respiratory problems, gynecological procedures, and to treat parasites. The Assyrians and Babylonians recommended honey to treat diseases of the ears and eyes. Even Hippocrates noted that honey cleaned wounds, helped to clean ulcers and sores, and was effective in treating ulcers on the lips, and healed carbuncles (Crane 1999).

One of the medical papyri from Egypt reported that honey mixed with red natron, northern salt, should be applied to improve the skin. Another cosmetic formulation in the Ebers papyrus recommended honey and turpentine as ingredients for a concoction used to wash the face would also remove wrinkles. The use of honey in cosmetics continued in Rome, where Nero's wife used honey mixed with asses' milk as a facial cream. So popular was this practice that it continued to be employed by women long after Nero's wife's death (Forbes 1965).

Honey was also used to make the incense kyphi, which was used in temples. The physician, Rufus who resided in Ephesus around 50 CE, recorded the recipe for kyphi. First, grind bdellium (an oily gum) with myrrh and wine until a thick liquid. Then add a mashed mixture of raisins and honey. To this add burnt resin, camel grass, sweet flag, Cyperus grass, saffron, spikenard, Aspalathus (a thorny plant used to thicken unguents), and cassia, then shape into pellets. The ingredients for kyphi were also recorded in hieroglyphs on the walls of Edfu Temple, Philae Temple, and were recorded in the Harris Papyrus. There were variations in the ingredient list for these various formulations, but honey was always included in the base (Manniche 1999).

These various uses of honey in antiquity made honey an important commodity for exchange. In Ancient Egypt, the demand for honey was so great that it was imported from neighboring countries. There are several examples in the tomb of Rekhmire where honey was brought to pay tribute to the Pharaoh. Ancient Egypt did not have a currency before the Ptolemaic and Roman periods. Instead, the economy ran on an exchange of goods based on their comparative value. To provide some consistency to a commodity's value, the Egyptians used a *deben* as a standard for valuation. One *deben* was a specific weight of copper, silver, or gold. For copper one *deben* would have been 91 grams of copper. However, the value was not in the metal itself. Rather it was based on the value of what could be fabricated with a *deben* of copper. For example, a copper vessel required 20 *debens*, which determined the value of the vessel.

Workers would not be compensated for their efforts in pieces of copper, they would receive the equivalent value of their wages in a commodity. If the wage for a day laborer was 2 *debens*, then the worker would receive 2 *debens* worth of grain. That grain, in turn, could be traded for other goods such as cloth, meat, bread, ducks, or honey.

Honey, being a liquid, would be dispensed in a unit of volume, the *hin*, to determine how much honey was valued at one *deben*. There are alabaster vessels from Ancient Egypt that state they contained one *hin*, which was on average 0.48 liters. Records from the Nineteenth Dynasty during the reign of Ramesses II can be used to calculate that one *hin* of honey was worth 1.2 *deben* of copper. Just like commodities today may fluctuate based on supply and demand, so did the value of honey in the ancient world. Honey dropped in value over the century after Ramesses II from 1.2 *deben* to 0.76 *deben*.

Of course, honey was not the only product that the bees provided the people of antiquity. Another product was beeswax. Beeswax, although not considered important by Columella in his writings, was the raw material for several needs. Aristotle believed that beeswax was produced by flowers and consisted of resins and plant gums (Fraser 1931). He also wrote that the bees carried the wax back to the hive on their legs, which suggests he thought that the pollen ball was wax. Pliny the Elder (XI.8.8) agreed with Aristotle, stating that "bees form wax from the blossoms of all trees and plants" (Hepburn 1986). In reality, the pollen ball is a mass of pollen grains that the forager bee forms by packing the sticky pollen in a depression on the hind legs called the pollen basket. We now know that beeswax is made by wax glands on the underside of the bee's abdomen. The wax is secreted as a liquid that solidifies into a wax flake. The bees get the energy to produce wax from the honey and pollen they eat. They use the flakes of beeswax to make their honeycomb within which the queen lays her eggs to produce the next generation of workers, and within which the hive bees place their enzyme inoculated nectar as it starts the transformation into honey. Beeswax does not dissolve in water, it does not discolor, and it can be carved or easily molded into shapes that may last for thousands of years.

The Greek physician Dioscorides, who lived during the first century, described how beeswax was cleaned after the honey had been removed. He wrote:

> How wax is made white as follows. Cut clean wax into small pieces, put it into a new jar and pour on it as much sea water (taken out of the deep) as shall be sufficient, and boil it, sprinkling a little saltpetre [potassium nitrate] on it. When it has boiled two or three times remove the jar, let it cool, take out the calce [lime] and scrape off the filth if there is any around it, and boil it again, adding more fresh seawater to it. When the wax has boiled again (as before) remove the jar from the fire, take the bottom of a new little jar (first moistened in cold water), let it down gently into the wax, dipping it in a little with a soft touch, that a little of it may be taken, and that it may be coalesced together separately. Having taken it up, pull off the first cake and let down the bottom of the jar again, cooling it in water again, and do this until you have taken up all the wax. Then pierce the little cakes with a linen thread and hang them up at some distance from one another. In the daytime set them in the sun, sprinkle them every now and then with water, and at night set them under the moon until they become perfectly white.
>
> (Forbes 1958)

Today, we use beeswax candles as a source of light, and that was also the case in antiquity. Beeswax burns at the relatively low temperatures of 143° to 151° F (60° to 66° C), and as it burns it gives off a very bright light and does not leave any ash. This was perceived as a magical property of beeswax, especially to the Ancient Egyptians and there are images of beeswax candles in New Kingdom paintings (Murray 2004; Kritsky 2015). The Etruscans (768–264 BCE) also made candles, but there is some debate as to whether they were made with beeswax. The Romans also burned rolled papyrus that was coated with wax for light (Forbes 1958).

It is difficult to pinpoint the origin of the beeswax candle. Candles were also made of tallow, which is made from rendered beef fat that is solid at room temperatures. Beeswax was much more expensive during antiquity resulting in a wider use of tallow candles. Both tallow and beeswax candles in antiquity were made by dipping a wick of twisted fibers from papyrus into melted wax or tallow and letting it cool. The continued dipping made the candle thicker, permitting it to burn longer. Due to its cost, the beeswax candle was initially a luxury of the wealthy. The British Museum collection includes a wax candle that was found in a Roman Era tomb in France from the first century (Forbes 1958; Crane 1978).

The oldest use of formed beeswax objects by the Egyptians date back to the Ninth to the Eleventh Dynasties. They carved wax figures of a deceased queen and placed it in a miniature coffin to function as a spare body for the dead (Kritsky 2015). Examples of these carvings can be found in the Metropolitan Museum of Art in New York, the Boston Museum of Fine Arts, and the British Museum in London. These wax objects were considered to function the same a shabtis, the small figurines that were thought necessary to carry out duties as needed as the deceased.

By Egypt's Third Intermediate Period, the burial practices had changed. No longer were the internal organs placed in canopic jars which were, in turn, stored within a box or an elaborate chest. Rather the stomach, liver, intestines, and lungs were removed during mummification, then embalmed, wrapped in linen, and placed back within the mummy's visceral cavity. The canopic jars, which were closed with stoppers in the form of the heads of the four sons of Horus, were replaced after the Twentieth Dynasty with carved beeswax figurines carved of the four sons of Horus. The sons of Horus provided a defensive magic for the deceased (Kritsky 2015).

Beeswax figures were also carved of animals including the *benu* bird, or phoenix, which were included with the mummy to guarantee the resurrection of the deceased. Other beeswax animals included scarabs, which were also placed within the mummy's linens to help with the rebirth of the dead.

Wax figures were also used with spells and charms in Egypt. For example, a carved hippopotamus was considered to be a spell to send a dream to a particular recipient. Other wax carvings were used to encourage more commerce or to encourage and "bind a lover."

An important use of wax involved is contribution to industry. The apiary at Tel Rehov may have included up to a hundred hives and, like beekeeping in Ancient Egypt, the beekeeping operation at Tel Rehov was a large and sophisticated operation to produce not only honey, but also wax. In fact, Mazar and his colleagues proposed that the production of wax may have been the primary importance of the operation at Tel Rehov. Beeswax during antiquity was a prized substance with many uses, but important among those uses was lost wax casting.

Lost wax casting, also called cirie perdue, is a method used to create fine complex objects such as votive figures and vases out of metal. The process involved carving a model of the desired object out of beeswax. Beeswax was the ideal medium for such models as it could be carved with fine details. The wax model would be placed in wet packed sand or plaster to form a mold of the model. Molten bronze, silver, gold, copper, or iron would be poured into the mold to cast the metal into the desired object (Hunt 1980).

The oldest objects made from the lost wax method were found in the region of the Dead Sea. In 1961, over four hundred copper objects were found under a straw mat, which carbon dated to 3500 BCE. The discovery of objects made by lost wax casting does not establish that beekeeping was practiced as the wax could be procured from honey hunting. However, the availability of large quantities of beeswax would be necessary to support large metal works operations, and it is suggested that this may have been the purpose for the large apiary at Tel Rehov. Beeswax was not the impetus for the origin of apiculture in Egypt as lost wax casting was not practiced in Egypt until the Eighteenth Dynasty over five hundred years after the carving of the Fifth Dynasty bas-relief mentioned earlier (Kritsky 2015).

Beeswax had other industrial uses in the ancient world. It was an important surface finish for Egyptian boats to protect the bright colors that were painted on the hulls as well as sealing them. Pliny wrote that the paint was applied with a brush using melted wax, and that "this process of painting ships is not spoilt by the action of the sun nor by salt water" (Crane 1999). Wax was also used as a binder for paint. A wooded box from the tomb of Tutankhamun included a coating of beeswax over the yellow paint to keep it from flaking away. Ramesses III 's granite sarcophagus in the Louvre in Paris was also treated with beeswax to keep colors attached to the incised hieroglyphs. Beeswax also served as a luting agent to fix alabaster lids to their vases, to affix a vase to a pedestal, and to set flint knives to handles (Lucas and Harris 2011). Recent chemical analyses have documented that beeswax was an important component of other ancient adhesives. The Etruscans used beeswax, birch bar tar, and pine resins to repair and to waterproof amphora (Rageot et al. 2015).

The malleable nature of beeswax made it ideal as a writing surface to record private information, details that would not warrant the use of papyrus or being carved on stone. As early as 1300 BCE, wooden tablets would be hinged together into book form. When opened, the facing "pages" would have recessed centers that would hold smooth beeswax which served as a writing surface. A pointed stylus would be used to record the information that could be later "erased" by making the wax smooth again. Pliny asserted that such tables were used during Homer's time (1250 BCE), and they continued to be used in Egypt during the Ptolemaic Period (300–30 BCE). The Roman poet Ovid documented their use during the last century before the Common Era when he wrote, "she holds the wax, begins writing and pauses, writes on the tablets and hates what she writes, both noting and erasing" (Enterline 2000) (Figure 4.6).

Beeswax also played a cosmetic purpose in antiquity. The Egyptians used beeswax to style wigs made of human hair and plant fibers. The wigs were styled with small tight corkscrew curls running the length of the hair, and these curls were held in place with beeswax. The styling would involve melting the wax and applying it to the curl while it was still liquid. As the wax cooled, the curls would set (Kritsky 2015).

Beeswax was also an ingredient in facial creams. The Egyptian recipe to keep the skin young in appearance was to grind resin, wax, Cyperus grass, and balanos (the desert date) oil together, and combine it with fermented juice. The resulting "cream" was to be used daily to reduce the appearance of wrinkles (Manniche 1999).

Beeswax was important in keeping unguents from separating. A goose fat unguent made in Egypt used goose fat that was concocted by bruising palm shoots, Cyperus grass, balsam wood, and sweet flag. This was combined with Lesbos wine then boiled and cooled three times. After the final boiling it was filtered using clean linen and allowed to set. Over time, the ingredients in unguent would separate during hot weather. However, the addition of beeswax to the mixture made the unguent much more stable (Manniche 1999).

Beeswax also played a role in preserving the dead. The Ancient Egyptians used beeswax to seal the nose, ears, and the incision used to remove the internal

FIGURE 4.6 A Roman-Era beeswax writing table. Photograph by Gene Kritsky.

organs. One female mummy embalmed during the Middle Kingdom had her back and thighs covered with a thin layer of beeswax. Herodotus reported that the Scythians preserved the corpse of their king with a covering of beeswax. This was also done in ancient Sparta where kings were covered with melted wax. A funeral practice in ancient Rome was to pour wax in a plaster life mask of a recently deceased individual to produce a wax effigy of the dead. According to Pliny these wax effigies would be set on a sideboard and carried during the funeral procession (Crane 1999).

Beeswax was a critical ingredient in creating the Fayum mummy portraits that date to Egypt's Roman Era (30 BCE–337 CE). These portraits were unlike any paintings produced in Egypt prior to the Ptolemaic Period. Pliny called the technique encaustic painting, from the Greek word meaning "to burn in." The portraits were painted on boards that were prepared by laying down a surface layer of plaster to receive the paint, which was made by melting beeswax then mixing in the pigments. The thick liquid or paste was then applied using wax modeling tools such as a graver (Forbes 1965). The results produced vivid colors and the thick layers of beeswax produced a sculptured surface that preserved facial details of the subject making them appear more lifelike that earlier Egyptian paintings. It is thought that the portraits were made during the subject's lifetime and were displayed in the home. Following the subject's death, the portrait would be trimmed to be incorporated into the mummy wrappings. The origin of encaustic painting is a mystery as the oldest known examples are Egypt's Fayum mummy portraits, but it is widely believed that it was developed by the Greeks (Figure 4.7).

The Metropolitan Museum of Art in New York has a Greek terracotta krater (350 BCE) with an image on the surface showing an artist painting in encaustic. He is holding a small cup with the heated wax and pigments in one hand and is applying the wax to the surface of a statue (MET 2017) (Figure 4.8).

Honey bees produce more honey than beeswax, and considering the various uses, beeswax became an important commodity in the ancient world. In Egypt, wax was used in their barter economy. Like honey, wax was valued in *debens* and one *deben* of beeswax was worth a half of a *deben* of copper (Kritsky 2015). Wax, along with honey, was also important for trade in the Mediterranean. Strabo wrote that the Iberian people Turdetania exported wax. Rome accepted wax in tribute (Francis et al. 2016). Unlike honey that was a viscous fluid, wax could be formed into cakes that could be easily transported as a solid provided the temperature did not get above 143° F.

Not all insect products in antiquity came from the hive. The ancient Egyptians' eye for color led them to appreciate the iridescent colors of buprestid beetles. Howard Carter reported that the tomb of King Tutankhamun contained "ceremonial canes, which are inlaid with elytra of iridescent colors" (Keimer 1938). Dyestuffs produced from insects colored the ancient world. When

FIGURE 4.7 A Fayoum encaustic mummy portrait, 100–150 CE. Metropolitan Museum of Art.

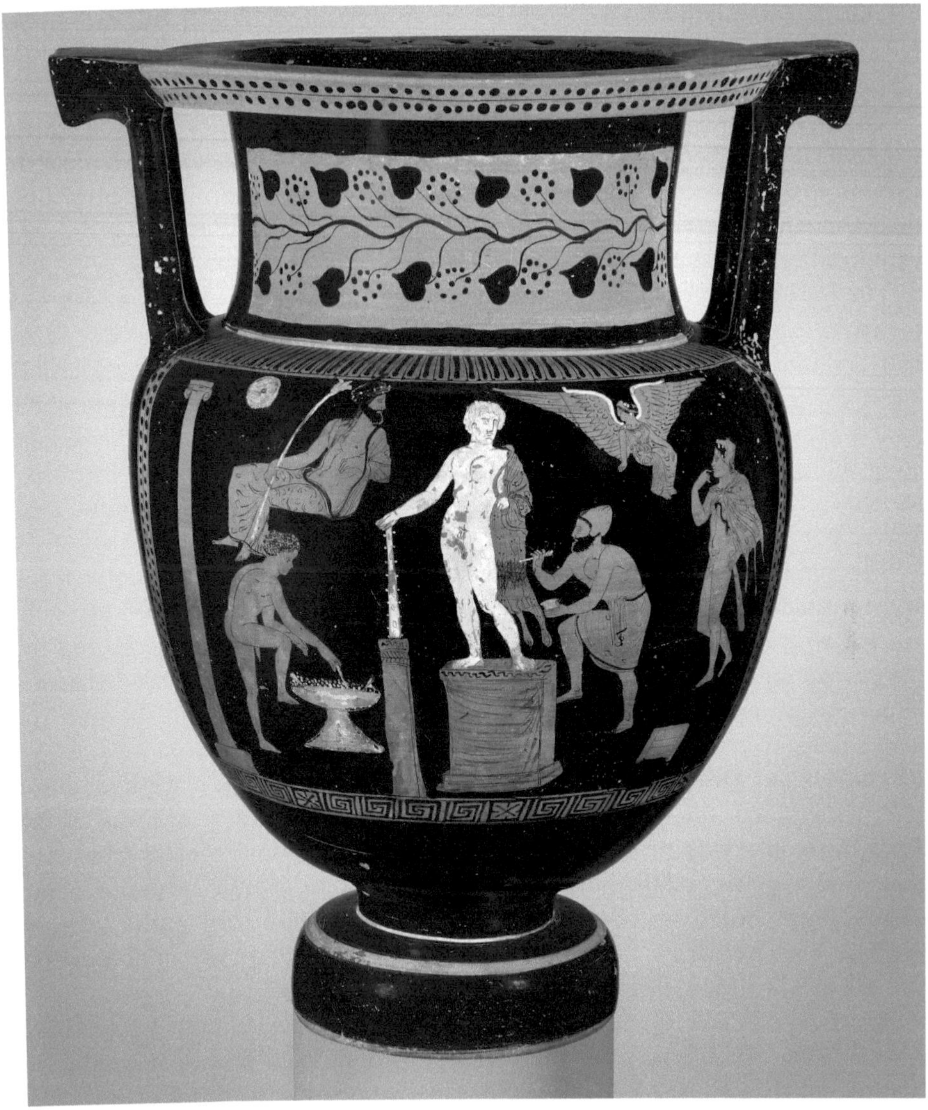

FIGURE 4.8 A Greek terracotta vase showing an artist painting using a beeswax binder, 360–350 BCE. Metropolitan Museum of Art.

Sargon II invaded Urartu, a mountainous area south and between the Black and Caspian seas, in 714 BCE, it was recorded that he took "scarlet textiles of Ararat and Kurkhi." It is believed that this cloth was dyed with the now endangered scale insect (*Porphyrophora hameli*) which was used to produce Armenian red, also called Armenian cochineal. The Armenian cochineal scale insect lives on the roots of several grasses. When mature, 10–12 mm long red females are collected, killed with vinegar fumes, dried, and crushed. The dried dyestuff was a valuable commodity for trade. Like the American cochineal, the Armenian red resulted from the presence of carmine acid that the insect uses as an anti-predator defense (Donkin 1977).

There are several references in the Old Testament to scarlet or crimson which would have been produced from a different scale insect, the kermes scale that lives on the evergreen oak (*Quercus coccifera*). This kermes red was prized by the Egyptians, and it is thought they obtained the dye from Phoenician traders who lived in Palestine where the evergreen oak grew (Donkin 1977).

The female kermes scale is a wingless sucking insect that forms a scale on the twigs of the evergreen oak. When it is mature it is a 6–7 mm globose scale that contains thousands of vivid red eggs. Approximately seventy thousand insects were required to produce one pound of dye (Brill 1980). The dye, when mixed with alum, causes a precipitate that binds with fabric giving it the scarlet color. Pliny wrote of the color, "red, as of the kermes-insect, which, from the lovieliness of the dark rose, shades, if you look up at in a bright light, into Tyrian purple, double-dyed purple and Laconian purple" (Forbes 1956).

The Sumerians, Phoenicians, and the Greeks used kermes red to color food and for medicinal purposes. Pliny reported that it would be ground up with vinegar to make a poultice to help heal wounds. Its value is underscored by the inscription of the great deeds of Tiglath-Pileser I, who recorded that he imported the evergreen oak into Assyria around 1100 BCE, well beyond its native range to establish a local source of the dye (Bucklow 2016).

The oak nut gall was recognized in antiquity as a source of black dye. Pliny (16.9) wrote that the oak gall nut could produce a black dye that was useful for dying wool. Galls are abnormal plant growths that result from larval insects that produce chemicals stimulating the plant to produce an excess of plant growth hormones that produce wart-like growths, or in the case of the oaks, to produce a round hard growth. English translations of Pliny's terminology for the oak gall include gall-nut or the oak-apple.

These insect-derived and insect-triggered dyes were used to color textiles, such as linen, wool, goat hair, and palm fibers. Fine textiles were a symbol of wealth and status in the ancient world. Linen and palm fibers are plant-based fibers, whereas wool and goat hair are derived from animals. However, there is one textile that is insect based, silk, which dramatically captured the attention of the ancient elite.

Pliny the Elder wrote in his *Natural History* Book XI.26 on insects:

> There is another class also of these insects produced in quite a different manner. These last spring from a grub of larger size, with two horns of very peculiar appearance. The larva then becomes a caterpillar, after which … it becomes a silk-worm. These insects weave webs similar to those of the spider, the material of which is used for making the more costly and

luxurious garments of females, ... which, while they cover a woman, at the same moment reveal her naked charms.

The silk that Pliny is describing is not considered true silk. True silk is produced by the domesticated moth larvae belonging to the species, *Bombyx mori*. Pliny was talking about the silkworm on the island of Cos, which is from the wild silkmoth (*Pachypasa otus*). Panagiotakopulu et al. (1997) described a silk cocoon that was found in an excavation on the Greek island of Santorini that dated over two millennia prior to Pliny. Although the identification of the silkmoth that produced this cocoon is not definitive, it appears it was most likely *Pachypasa otus*. However, the presence of a silken cocoon on Santorini does not establish that silk cloth was being produced. Indeed, a critical examination of all the pre-Roman "silks" found in Europe have brought into question these interpretations as being evidence of true silk (Jørgensen 2013).

What is known is that silk was imported into the Mediterranean following the alliance of King Mithridates II of the Parthian Empire with the Han Emperor of China Wu Ti, which was in place by 120 BCE. This opened the Silk Road to trade with China and domesticated silk started being imported into Rome. The Parthian Empire occupied today's Iran and Iraq from 247 BCE to 224 CE. Within a century of the alliance, Julius Caesar had silk curtains, and the silk trade dramatically increased during the reign of Augustus (Thorley 1971).

The Silk Road trade was not a direct distribution from China to Rome. By the time imported silk passed through the hands of various intermediaries (each of whom made a profit) it became very expensive. The silk that arrived in Partha was not the thin delicate cloth that was so prized by the Romans, but rather a heavy closely woven cloth. The enterprising Parthians would send the silk to Eyre, Sidon, Bertyus, and other towns where the silk was unraveled and rewoven into the delicate cloth that was in demand. This cloth was so unique that the Parthians, who did not possess this atural resource that was in demand, were even selling their fine silk cloth back to China (Thorley 1971).

CONCLUSION

The ancients' use of insect products evolved during the 3,500 years of antiquity. Beginning with the simple robbing of honey from wild bee nests, throughout the ancient world people were discovering how to encourage honey bees to take up residence within tubes of dried mud, elongated pottery jugs, hollowed out logs, boxes made of sticks or boards of wood, and eventually brick niches in walls. This association of humans with honey bees was as valuable to the ancient world as was the domesticated dog, cow, and goat.

The development of beekeeping provided honey and beeswax. The caloric-rich honey was not only a food for the elite, but an ingredient in other recipes as humans developed cuisines that characterized their cultures. Honey proved to be an important medicinal as different cultures, through trial and error and simple luck, relied on honey to stop infections and to heal cuts and burns millennia before we knew there were bacteria.

Beeswax, at first a by-product of the desire for honey, was an insect product that glued flint blades into handles, repaired containers for continued use, illuminated temples and palaces, and was used to make people more attractive. Its use in lost wax casting permitted the production of bronze, copper, and iron objects to be used for construction, warfare, religious rituals, and to adorn the dead for the afterlife.

The discovery of insect dyes documents the curiosity of the ancient mind as people scraped away scales of insects and processed them to dye garments. Fashion owes a debt to the insects that provided the crimson color for silk and wool clothing.

This was the world that the people of antiquity created and left for the Middle Ages. With this historical foundation, insect products would contribute to the development of new economies, and new political and religious institutions. Subsequent history will show us that the insect products of antiquity would endure in their importance long after the ancient empires crumbled into ruin, becoming tourist attractions.

CHAPTER FIVE

Insects in Mythology and Religion

RON CHERRY

For more than forty thousand years before European navigators visited the shores of the Great South Land, Aborigines occupied Australia, including its arid deserts, tropical rainforests, coastal plains, mountains, and especially its major river systems. Estimates by anthropologists put the population of Australian Aborigines before 1770 at more than three hundred thousand. They spoke five hundred different languages grouped in thirty-one related language families. Aborigines were completely at home in their surroundings and had no trouble "living off the land." This was mainly because of their intimate knowledge of the topography and natural resources of their tribal territories, and their complete understanding of the habits of the animals they hunted. The Australian Aborigines extensively used insects from their surroundings as food, medicine, and as part of their cultural beliefs. Insects have been consumed as food in many parts of the world, and insects consumed directly as food was probably the most important use of insects to the Australian Aborigines. Insects used for food were moths (Noctuidae), ants (Formicidae), lerp insects (Psyllidae), honey bee hives (Apidae), termites (Termitidae), and grubs (Cossidae). Cockroaches (Blattidae) were used for a local anesthetic. Ants (Formicidae) were used to cure headaches, as a cold remedy, antiseptic, and expectorant. Several other insects were also used for medicinal purposes. And interestingly, the didgeridoo (wind instrument) was made from tree limbs hollowed out by termites (Termitidae) (Cherry 1991).

INSECTS IN THE BIBLE

One of the earliest books containing numerous references to insects is the Bible. Perhaps no other single book in the western world is so frequently cited as the Bible. It is used to point a moral, to adorn a tale, to win an argument, and to title a book. This all-time bestseller is also a rich source of quotable material on insects and their relatives—their record of devastation; their use by philosophers to exemplify desirable attributes of industry and humility; their ravages pointed to as retribution for evil doing (Bruce 1958).

An early publication by Wood (1869) attempted to determine specific insects in the Bible. However, he found this difficult since insects could not be accurately identified. A later attempt to identify the insects was made by Dow (1917, 1918) who found the King James Version especially confusing. This is consistent with Birdsong (1934) and Montgomery (1959) who also found much confusion. For example, they found nine different words for locusts mistaken for other insects.

Kritsky (1997) in a more recent review notes that the King James Version of the Bible was not based on original texts, but on earlier translations. Moreover, translators working on the King James Version invented new English words including for insects. However, there have been more recent developments in biblical scholarship. The Dead Sea Scrolls were translated and newer English translations of the Bible based on original texts have been made. With this in mind, Kritsky (1997) published a new revised version of the insects and other arthropods of the Bible. This revision removes several orders of insects from the Bible, adds new groups. Also, the revision reduces insect and arthropod references to 98 from the 120 found in the earlier King James Version (Bruce 1958).

The best-known examples of a god punishing people with insects are found in the Bible being plagues to punish early Egyptians. This topic was reported very early in entomological publications by Dow (1917, 1918). It has been a popular topic since then, reported in numerous entomological works. Recent discussions of this topic may be found in Kritsky and Cherry (2000) and Carde and Resh (2012). Essentially the story is as follows. The god Jehovah forced the Egyptians to release the Israelites from slavery by subjecting the Egyptians to ten great plagues. As many as five of the ten plagues in the Bible are directly or indirectly attributed to insects such as flies, lice, murain, locusts, boils, and possibly darkness (Berenbaum 1995).

Later in the Bible, locusts again appear in Deuteronomy as curses for disobedience to God (Horne 1908). And yet again locusts appear in the Bible in an even more horrific form at the predicted Apocalypse at the end of the world. In appearance, the locusts are like horses equipped for battle. Their faces are human like and they have teeth like lions. Their tails are like scorpions with stingers. They are not to damage the grass of the earth or any green growth, but only those people who do not have the seal of God on their foreheads. The locusts torture people for five months, but do not kill them. People will long to die, but will not die (Verdet 1987).

Similar examples of locusts being used as divine punishment are found in other non-biblical sources. In Chinese mythology, different hells exist to punish sinners for different sins. The sixth hell is ruled by Pien ch'eng who punishes those guilty of sacrilege. There are several variations of punishment here, one of which is being devoured by locusts (Mercatante 1978). Across the world in the pre-Columbian mythology of the Toltec of Mexico, it is reported that their king, Huemac the Wicked, disgusted the gods so much that the people were plagued for six years including cloud-like swarms of locusts (Spence 2005). Similarly, in the pre-Columbian mythology of the Aztecs of Mexico, a man is transformed into a locust called *ahuacacha-pullin* for transgressions against the gods (Mackenzie 1923). An extensive review of insects and divine intervention is found in Cherry (2015).

INSECTS IN GREEK MYTHOLOGY

Besides the Bible, another source of early Western literature is the recorded stories of Greek mythology. The birthplace of western civilization, the ancient Greek empire flourished from the eighth century BCE until its conquest by the Romans in 146 BCE. Because the Greeks placed such a high value on the written word, many of their most ancient myths have survived almost completely intact. The Greeks were the first to dramatize stories for performance on the stage, and also gave us the words we still use to describe epic and universal tales—myths from Greek work mythos (Pemberton 2015).

The *Iliad* and the *Odyssey* by Homer contain the oldest Greek writings we have. There is no way to accurately date any part of them. Scholars differ widely and will no doubt continue to do so. An unobjectionable date as any is 1000 BCE—at any rate for the *Iliad*, the older of the two poems. A second great Greek mythologist was Hesiod being placed in the eighth or ninth century BCE. Hesiod was a poor farmer whose life was hard and bitter. But Hesiod had much to say about the gods in writing the poem *Theogony* which was entirely concerned with mythology. The *Theogony* is an account of the creation of the universe and is very important for mythology (Hamilton 1942). Aesop's Fables are a collection of fables credited to Aesop, a slave and storyteller who lived in ancient Greece approximately 620 to 564 BCE. Although initially an oral tradition, these stories were written down after Aesop's death. Insects are found in many of these fables as tricksters or moral guides. Other later Greek writers of Greek mythology were Pindar, Aeschylus, Sophocles, and Euripides. Later, of course, much of Greek mythology was adopted into Roman mythology. Various Roman writers continued to record the stories of Greek mythology. However, to Romans the stories were infinitely remote, mere shadows. Hence, the best guides to a knowledge of Greek mythology are the Greek writers, who believed in what they wrote (Hamilton 1942).

Numerous references to insects in Greek mythology are scattered throughout entomological and mythological sources. However, the single best reference to this topic by far is the book *Greek Insects* by Davies and Kathirithamby (1986). The following are just a few of the numerous examples of insects in Greek mythology given in the book (Figure 5.1).

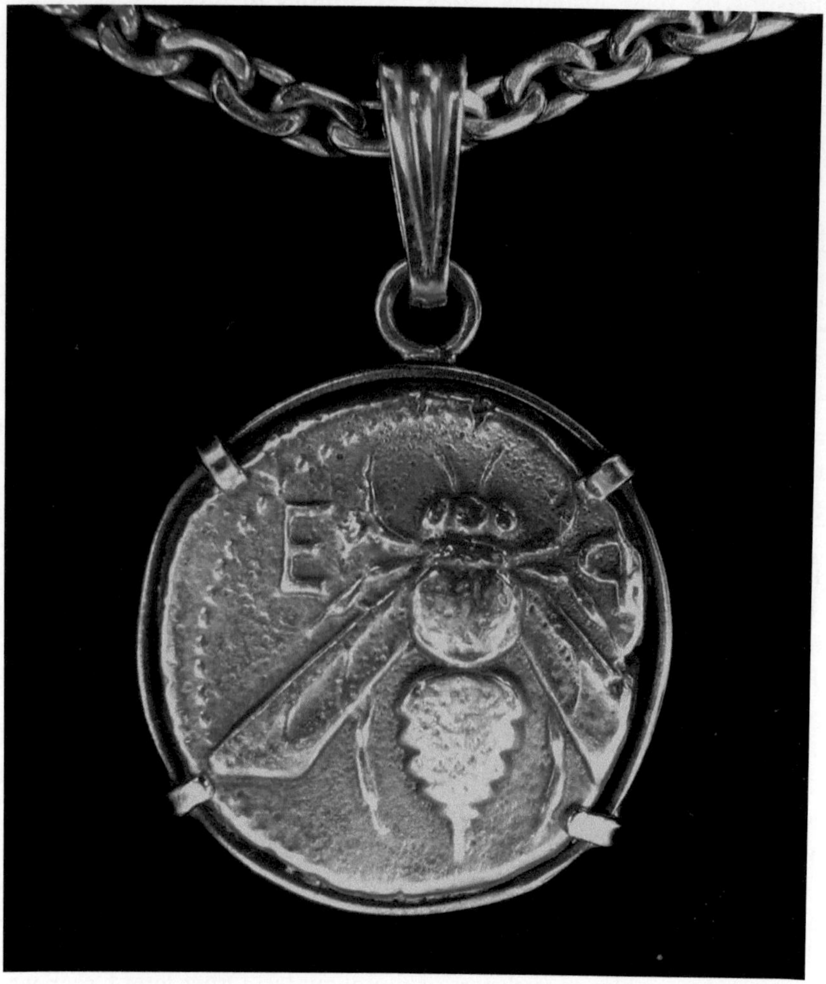

FIGURE 5.1 Ionian coin showing the bee which was a cult symbol for Artemis, the goddess of wild nature. Photograph by Ron Cherry.

Three examples of insect mythology being used to explain natural phenomena involve the great Greek god Zeus. In the first, a dung beetle takes revenge on an eagle, the bird of Zeus, by destroying its nest. The eagle seeks shelter by laying its eggs in the lap of Zeus. Thereafter, the beetle flies up to Olympus and deposits dung in the god's lap. When the god leaps up to brush the dung away, the eggs roll from his lap and are broken. This nature myth has been used to explain why eagles lay their eggs at a time of the year when no beetles are around. In another story, a lazy, greedy farmer steals fruits of his neighbors. Zeus punishes him by changing him into an ant, but not his disposition since the ants still collect wheat and barley of others for their own use. And in another story, bees asked Zeus to make their stings mortal to men who stole their honey. Zeus punishes them for their malice by decreeing that the bees themselves should die after using their sting. The common theme running through the previous myths is that they explain how the world came to be as it is. One important function of mythology in general is to explain various features of life. How did my people get fire? Why does a zebra have stripes? How did ants come to collect wheat and barley? Why should I be good? Where will I go when I die? Explaining our world and other functions that insects have served in mythology are found in Cherry (2002).

The Greek concept of the Lepidoptera is especially interesting. The Greek name for them is "psyche," the word for the soul. The adult's emergence from the darkness of the pupa called in Greek "nekydallos," or little corpse, is presumably the key to the metaphor in conjunction with the popularity of the concept of winged souls. Interestingly, several writers have noted the strange paucity of references in ancient literature to these insects which more modern times have found so attractive or symbolic. One explanation for this is that the Greeks found something ominous or uncanny in butterflies as something not to be lightly spoken of. In contrast, visual artists seem to have labored under no such inhibitions as to the depiction of the creature and often exploited the opportunities for symbolism offered by its ambiguous name (Davies and Kathirithamby 1986).

Bierlein (1994) notes that the famous Greek myth of Cupid and Psyche is, on the surface, a charming love story with a moral. But it makes a great deal more sense when it is understood that the names of the characters have meanings. Cupid is the Roman name for the Greek god of love, Eros: "eros" means sexual love in Greek. Psyche means both soul and butterfly in Greek. Therefore, this myth is a statement on the relationship between physical love and soul love, wherein the soul, like a butterfly, undergoes a metamorphosis.

The intervention of gods with insects in human affairs essentially falls into two categories. First, gods punish people with insects for bad behavior. Second, gods are prayed to by people to reduce insect problems. Some examples are better known to entomologists such as insect plagues in the Bible. Other less known examples are scattered throughout entomological literature (Cherry 2015). Several of these interventions are found in Greek mythology. The goddess Hera sent a gadfly to harass Heracles in one myth (Ions 1999). Hera also sent a gadfly to harass Io in another myth (Grimal 1965: 97–177). Bellerophon is stung by a gadfly while riding the winged horse Pegasus (Jobes 1962). In contrast to punishment, Greek gods were prayed to to reduce insect problems. In Greek mythology, the expulsion of flies was credited to Zeus, Heracles, or Apollo (Morge 1973) and Zeus was nicknamed "fly catcher" (Waldbauer 2012). Pliny mentions that during a festival to honor Apollo, an ox was sacrificed to flies and a god of flies was invoked for relief from annoying flies (Clausen 1954). The topic of insects and divine intervention is discussed in greater detail in Cherry (2015).

INSECT NAMES FROM CLASSICAL MYTHOLOGY

Much of Roman mythology was based on Greek mythology with Zeus becoming Jupiter, etc. The merging of these two mythologies became known as Classical Mythology. Interestingly, many insect names have been derived from Greek and Roman mythology (Cherry 1997). The practice of naming plants and animals based on mythological origins was also practiced by early naturalists. This practice was prompted by the classical education in Greek and Roman mythology received by many early naturalists (Yancey 1952). However, the use of mythological names for biological organisms has also been criticized. Schenk and McMasters (1956) opined that the chief virtue of mythological names in taxonomy was that they afford variety but lack descriptive value. In addition, Jaeger (1978) believed that mythological names were objectionable not only because they have been used so frequently in other classes that they are likely to be preoccupied but also because of ambiguity in application. For example, Atlas and Hercules may seem appropriate names when describing very large insects, such as the atlas moth (*Attacus atlas* (L.)) and Hercules moth (*Cosinocera hercules* (Miskin)). However, it is less clear why the insect genus *Pieris* was named for Pieris, a daughter of Zeus (Yancey 1952). And, of course, there is the chance that an organism with a mythological name may not have been named after a mythological figure. The alpine argus (*Albulina orbitulus* DePrunner) and brown argus (*Aricia agestis* Denis & Schiffermuller) are butterflies in the family Lycaenidae (Carter 1992). Unless one knows for sure, it is impossible to determine if the Argus in the common name was named after the ever watchful, multi-eyed monster of mythology (Hamilton 1942) or after a friend, or simply because someone liked the sound of the name.

Regardless of one's own viewpoint on naming biological organisms after mythological figures, these names do occur frequently in the zoological sciences including entomology. The atlas moth, hercules beetle (*Dynastes hercules* (L.)), and hercules moth were named after mythological figures of great size and strength. Atlas was one of the Titans who warred against the gods and was condemned to uphold the heavens on his shoulders. Due to an error by Rumold Mercator in 1595 when publishing a map collection, Atlas was depicted holding a globe (earth) rather than the sky on his shoulders; from Mercator's mistake, the word Atlas has been used for map collections (Macrone 1992). And, of course, Hercules was the mighty Greek hero who sailed with Jason and the Argonauts (Hamilton 1942) and, ultimately, became worshipped as the god of physical strength.

Another group of insects with an interesting common name derived from mythology is the amazon ants, who were named after the fierce female warriors, the Amazons, of Greek mythology (Mason 1974). These ants, which are comprised of five species in the genus *Polygerus*, are slave makers. Amazon ants are nowhere very common, but their striking appearance (large size, bright red or black coloration, and shiny body surface), the extraordinary degree of their behavioral specialization, and the spectacular qualities of their slave raids have placed them among the most frequently studied of all the ants (Holldobler and Wilson 1990). The workers are "femme fatales," possessing long, pointed mandibles for fighting. The analogy of these ants to the fierce female warriors of Greek mythology is apparent. These warlike women lived on the coast of the Black Sea where they formed a state from which men were excluded. To reproduce, Amazons ventured into nearby settlements where they used men to satisfy their sexual needs before heading home to await the results. Male babies were returned to the fathers or killed. Female babies were saved and raised in a strict regimen including burning off the right breast so it would not interfere with the handling of a bow or javelin (Macrone 1992). With the exception of the breast mutilation, the comparison of *Polygerus* ants to Amazons is appropriate because both groups are female-based societies sustained by ferocity and occasional raids.

Two other insects whose common names have been derived from well-known mythological figures are the polyphemus moth (*Antheraea polyphemus* (Cramer)) and the promethea moth (*Callosamia promethea* (Drury)). In the polyphemus moth, the large eyespots on the hind wings are the mythological connection (Mitchell and Zim 1962). Polyphemus was one of the Cyclops (giants with one large eye in the middle of their foreheads) who, occasionally, fed on human flesh. In Homer's *The Odyssey*, Polyphemus eventually was blinded by the Greek hero Odysseus after Odysseus and his crew were taken prisoner by this one-eyed giant (Hamilton 1942). The best-known legend about

Prometheus also contains a good dose of violence. Prometheus was the Titan who stole fire from the gods and gave it to humans. He later was punished by Jupiter, who chained him to a mountain where an eagle preyed upon his liver which was renewed as fast as it was devoured. Fortunately, Prometheus finally was released by Hercules (Bulfinch 1991). The association of the myth with the promethea moth is not obvious. Because of his exploits, the name Prometheus over time became synonymous with independence and daring. The cocoon of the promethea moth is suspended rather freely by silken thread. Hence, it is this developmental feature of the insect from which the moth has gained its elegant name (Yancey 1952).

Insect genera also abound in Greek and Roman mythological derivations. An example is the naming of the genus *Colias*. Colias is a lesser-known name for Aphrodite, the Greek goddess of love and beauty, and the name was given in reference to the beauty of the butterflies in the genus (Yancey 1952). Other examples are *Atropos* and *Clothilla*, named after the two Fates Atropos and Clothos (Sharp 1895).

Family names of insects have also been derived from genera named after Greek and Roman mythological figures. An interesting example concerns the psocopteran family Atropidae (=Trogiidae), the name of which is derived from *Atropos*. In mythology, Atropos was one of the three Fates who were goddesses who determined the course of human life; her role was to cut the thread of life (Bulfinch 1991). The genus *Atropos* was named after the mythological Atropos because it was thought that members of this genus made a tapping noise on wood at the death of humans (Sharp 1895). Females of some species in this family do produce a sound by tapping their abdomens on a substrate (Borror et al. 1992). However, it has been suggested that original "death watch" insects were not atropids but, rather, anobiid beetles that produce sounds in their galleries (Sharp 1895).

Another example is that of the family Saturniidae, the name of which is derived from the genus *Saturnia* (Swan and Papp 1972). Saturn was a major, well-known god of Roman mythology who, in an infamous act, devoured his children except Jupiter, Neptune, and Pluto (Bulfinch 1991). However, what makes the family interesting from a mythological perspective is that it contains some of the best-known examples of insects named after mythological figures. These insects include the atlas moth, hercules moth, io moth (*Automeris io* (F.)), polyphemus moth, and promethea moth. As a final note, Atlas and Prometheus were brothers in mythology. Prometheus consoles Princess Io by telling her that her race will one day yield a glorious hero. Io's descendant would be Hercules, one of the greatest mythological heroes, who ultimately freed Prometheus (Hamilton 1942). It seems poetic justice that through common names of insects, Atlas, Hercules, Io, and Prometheus have been re-joined again in the same family, the Saturniidae.

INSECTS IN NATIVE AMERICAN MYTHOLOGY

Numerous accounts of insect mythology are found scattered in early western literature. However, an abundant treasure trove of insect mythology was waiting to be found with the later conquest of the New World by Europeans. Although rarely written, Native Americans had rich and sometimes complex mythologies at the time of the European conquest. These mythologies were very old, although exact dates of origin cannot be determined. As Pemberton (2015) notes, since the vast majority of myths were passed down orally before the widespread adoption of writing, many myths are thousands of years older than written sources. Native American culture is rich in mythology, including numerous examples of animals as mythological figures. Powerful animal gods such as the plumed serpent, Quetzalcoatl, of Central America and the Thunderbird of the North American plains are prime examples. Large mammals such as the bear, buffalo, jaguar, and wolf are also frequently assigned mythological properties by various Native American cultures. However, although lesser known, insects also occur frequently in Native American mythology in roles ranging from trivial to cosmology. This is especially true of Native Americans inhabiting tropical areas, probably because of the richness of insects in their surroundings (Hogue 1987).

An interesting example of insect behavior explained by Native American mythology concerns the genus *Eleodes* of the family Tenebrionidae (Figure 5.2). As one approaches an adult *Eleodes*, it will lower its head to the ground, raise its abdomen, and emit a disagreeable odor as a defense against enemies. The Cochiti of the southwestern United States associated this behavior with another creation story. This story noted that long, long ago this beetle was assigned to place the stars in the sky. Unfortunately, the beetle became careless and dropped the stars that then scattered to form the Milky Way. So ashamed was the beetle for what he had done, that even today the beetle hides its face in the dirt when anyone approaches. Tenebrionid beetles also play in a minor role in the mythology of the California Youlmani Yokuts (Essig 1931). Another coleopteran that appears frequently in Native American mythology is the firefly (family Lampyridae) because of its bioluminescence. Fireflies play a role in a myth of the Mayans of Central America. In a story reminiscent of tales told in Homer's Odyssey, Mayan twins must keep two cigars lit all night. The twins appear to complete this Herculean task by substituting fireflies at their cigar tips, thus tricking their captors (Tedlock 1985). Among the Aztecs of Mexico, simple-minded people believed that fireflies were witches or wizards that went around at night throwing out fire from their head or mouth (Curran 1937). According to the Jicarilla Apaches of New Mexico, fire came from a mythical campfire ignited by fireflies (Frazier 1930).

FIGURE 5.2 According to Cochiti mythology, an *Eleodes* beetle created the Milky Way. Photograph by US National Parks. Public domain.

Mosquitoes are central to more stories than any other group of dipterans. Many of these stories are derived from the blood-sucking habit of the insects. In Mayan mythology, a mosquito acts as a spy by biting various lords to learn their names and discern them from mere mannikins (Tedlock 1985). Clausen (1954) noted several examples of mosquitoes in the mythology of Native North Americans. In one interesting example, the Tahltan of British Columbia account for a wood-boring beetle with the following myth: a long time ago, Wormwood (the larva of a beetle) and Mosquito lived together. Day after day, Wormwood saw Mosquito come home swollen with blood that he had eaten. When questioned, Mosquito, not wishing to give away his secret, replied that he had sucked it out of trees. Wormwood immediately attacked the trees, and to this day he and his descendants bore into wood looking for blood. Hitchcock (1962) reported that mosquitoes were the most prevalent pest in the far north of North America and described legends concerning the origin of these insects. Among Native Americans of the Pacific Northwest, a tale common to most of the groups explained the creation of mosquitoes from the ashes of a cannibal who was burned by the youngest of five brothers in revenge for killing his family (Bancroft-Hunt 1979).

Dipterans other than mosquitoes are also found in Native American mythology. Among the Hopi of the southwestern United States, kachinas are friendly spirits who are counterparts for the real world. These kachinas are personified into carved figures called Tihu by the Hopi (and incorrectly called "kachina dolls" by people unaware of the religious significance of the figures to the Hopi). The Hopi personify insect spirits of the Assassin Fly in various kachina figures (Wright 1977). The Auraucanian of South America believed that departed tribesmen, particularly chiefs, took the form of horseflies. These insects are regarded as spirits from the beyond, and horseflies appearing at feasts are accepted as an indication that the dead tribesmen are taking part in the feast (Clausen 1954). In Navaho myths, there is a big fly that will sometimes fly down and sit on the shoulder of one who is walking alone in the desert. This fly is known as Big Fly and whispers to young heroes the answers to questions that their fathers put to them when being tested. Big Fly is the voice of the holy spirit revealing hidden wisdom (Campbell 1988). The Montagnais of eastern Canada believed that the overlord of fish, particularly salmon and cod, was Big Biter (= *Tabanus affinis* Kirby). This fly appeared whenever fish were being taken from the water and hovered over the fisherman to see how his subjects were being treated. Occasionally Big Biter would bite the fisherman to remind him that the fish were in his custody and to warn him against wastefulness (Speck 1935).

Numerous legends also exist about hymenopterans. In Navaho mythology, ants are important actors in the Emergence Myth, and a whole chant complex, Red Antway, takes its name from the insect (Wyman and Bailey 1964). One interesting legend of the Yurok of California explains that ants and wasps came to be venomous by catching chips in their mouths from a mythical burning arrowhead (Waterman 1920). In another story, the Algonquin of Canada explain how bees, wasps, and hornets gained the power to sting. Ages ago, bees were industrious but could not sting. Hence, other animals stole their honey and bees frequently starved. After appealing to the Great Spirit, bees were given barbed stingers to protect themselves. Because the wasps and hornets claimed to be cousins of the bees, the Great Spirit also gave them stinging weapons. In another legend, the theme of industrious bees occurs again with moral overtones. In this story, two tribes of little people lived near each other. The difference between them was that one tribe looked for food and conserved it wisely while the other tribe played and sang and danced all day. Eventually, the Great Spirit reacted by making the industrious tribe into bees and the lazy tribe into flies (Clausen 1954). Hence, bees fly from flower to flower eating honey while flies eat food that has been thrown out, apparently a justified fate for lazy people. The concept that bees were industrious is also illustrated in Mayan mythology. Mysterious Mayan ceremonies connected with bees seem to have symbolized people of great industry and potential (Nicholson 1985). Hymenoptera also play minor roles in other Mayan legends (Tedlock 1985).

Because of the beauty, power of flight, and complete metamorphosis found in butterflies and moths, these insects are frequently found in Native American myths. In North America, the Blackfeet believed that a butterfly brought sleep and dreams. A Blackfoot butterfly symbol, a design roughly in the shape of a Maltese cross, was embroidered into buckskin and tied into a baby's hair to hasten the child's sleep (Grinnell 1899). The butterfly was also a prominent figure in the myth and ritual of the Hopi. This insect occurs frequently on prehistoric pottery, in a ritual "Butterfly Dance," and a clan in one of the Hopi pueblos was even called the Butterfly Clan (Fewkes 1910). The spirit of the butterfly is personified in Hopi kachina figures (Wright 1977). Born out of the caterpillar in the chrysalis, butterflies were a symbol of rebirth, regeneration, happiness, and joy to Native Americans in Mexico. In one legend, the powerful, plumed serpent god Quetzalcoatl first enters the world in the shape of a chrysalis, out of which the god painfully emerges into the full light of perfection symbolized by the butterfly. The Obsidian butterfly, Itzpapalotl, is a patron deity in the Aztec calendar (Nicholson 1985) (Figure 5.3). Native

FIGURE 5.3 Itzpapalotl, the "obsidian butterfly" was an Aztec goddess of childbirth. Photograph by Gene Kritsky.

South Americans also integrated various Lepidoptera into their mythologies (Hitchcock 1962). To the Goajiro of Columbia, if a particular large, white moth is found in a bedroom, it must not be mistreated for it is the spirit of an ancestor come to visit. If the moth becomes troublesome, it can be removed only with the greatest of care or the spirit may take vengeance. Among the Aymara of Bolivia, a certain rare nocturnal moth was thought to be an omen of death.

As noted in previous examples, insects played several different roles in Native American mythology. At one extreme, some insects were considered the revenge of demons upon mankind. At the other extreme, a butterfly spirit could gently bring sleep and dreams to a child. Insects in myths taught morality lessons, explained the physical environment, and, on a grand scale, even gave understanding to the origins of the earth itself. Unquestionably, insect lore provides much insight into how Native Americans viewed themselves and their universe (Cherry 1993).

CHAPTER SIX

Insects as Symbols

JAMES N. HOGUE

The depiction and transmission of ideas through the use of symbols is a human cultural construct that predates our earliest civilizations. Although a product of our distant past, the use of symbolism flourished with the advent of greater socialization and the apparent need to preserve and pass on ideology and history among peoples over space and through time. The study of symbolism in early human culture is anthropologically informative, especially with respect to the evolution of early cultures, by providing insight into the elements of nature that influenced the minds and thus the cultural evolution of ancient peoples.

Symbols are terms or pictorial depictions of things that may be familiar in daily life that convey additional meaning to that of their conventional meaning (Jung 1964). Sometimes this symbolic meaning is dependent on the context of the symbol's depiction. Symbols are typically used to represent concepts that are difficult to understand or define. Humans use symbols to convey human feelings, as instructional tools, as entertainment, and to solidify human associations (or disassociations) with nature. Sometimes they are used to validate or explain human behavior. People also generate and use symbols unconsciously during their dreams. Although usually clear to the user, symbolic significance is sometimes subtle or unknown to the recipient/viewer.

Because insects are ubiquitous and apparent elements of the historical and current experience of human beings, it is not surprising that they have been used across many cultures and times as symbols (Hogue 1975). The mythology of human cultures is, in itself, a form of symbolism (Kritsky and Cherry 2000), therefore a good place to search for the use of insects as symbols is within the written and orally transmitted myths, folklore, and legendary literature produced by ancient cultures that have survived to the present. Although often not used symbolically, depictions of insects in various forms of art can be another fruitful source of examples (Klein 2007).

Information on symbolism used by the peoples of antiquity is known to us through artifacts, art, and the written and oral transmission of myths, folklore, religious beliefs, and cultural history (e.g., Motte-Florac and Thomas 2003). Such examples incorporate insects, and other terrestrial arthropods that people often equated with insects, in these endeavors for a variety of purposes which shed light on human culture and are of interest to entomologists and anthropologists alike.

This chapter discusses instances of insect-based symbolism in antiquity, including the earliest periods of recorded human history. Most information on this subject comes from classical antiquity, involving the ancient civilizations around the Mediterranean Sea but will also include information from other peoples around the world representing their earliest cultural contributions to insect symbolism, that is from the time of their cultural antiquity. This discussion is organized topically, rather than geographically or temporally. Although the following topics are rather loosely constrained; for example, some symbols fit into more than one category and some treatments do not fit well into the chosen category, this arrangement facilitates the discussion by providing some measure of organization that reduces repetition that would result from other categorization schemes.

AETIOLOGICAL, INSTRUCTIONAL, AND DESCRIPTIVE SYMBOLISM

Some folklore is aetiological in nature, serving to explain something observed in nature and provide a reason for its occurrence. For example, the Tlingit people of coastal, northwestern North America have an aetiological story for the origin and characteristics of mosquitoes (Erdoes and Ortiz 1984). The ancient Greeks provide a mythological explanation (the actions of an angered Zeus) for the death of honey bees resulting from their stinging someone (Davies and Kathirithamby 1986). Insect-focused aetiological stories generally incorporate little symbolism on the part of the insect, but stories using insects to illustrate broader concepts often do. This use of insect symbolism is often in the form of analogy and metaphor.

The comparison between colonies of eusocial insects, especially the familiar honey bees and ants, and the activities of humans is ancient and widespread (Farrell 1985). Writers and storytellers used insects such as these which had such well-known attributes to convey particular messages or teach a lesson. Among the most well-known purveyors of this in antiquity was the Greek fabulist Aesop. Aesop was a storyteller living on the Greek island of Samos during the early part of the sixth century BCE. Originally from the ancient Greek city of Mesembria in Thrace (now eastern Bulgaria), he was a slave under Iadmon, who later freed him. Aesop's fables, disseminated orally at the time, are well known for their humor and power to enlighten the mind. These fables, relying heavily on metaphor, inherently involve symbolism.

Aesop used a wide variety of insects in his stories including ants, bees (and honey), beetles, butterflies (and caterpillars), cicadas (often erroneously portrayed as a cricket or grasshopper), fleas, flies (including gnats and mosquitoes), lice, scorpions, spiders, and wasps. In most cases, these are not mere actors, but are chosen to produce or emphasize the salient point. Foreknowledge of a particular insect's inherent characteristics lends these creatures to symbolic usage in verse and tale. Although the ancient Greeks generally considered such small creatures as rather insignificant and unworthy of discourse (Davies and Kathirithamby 1986), this very quality was employed by Aesop in his instructive tales of the triumph of the small over the great or the virtues of hard work and diligence. In fact, nowhere else do insects find such an important role in Greek literature than in the enlightening tales of Aesop. A good example is the tale of the ant and the cicada.

The story takes place in a field as winter is approaching and food is becoming scarce. The second player in the story is often portrayed as a grasshopper or cricket, but the Greek word *tettix* as used here refers to a cicada (Egan 1994). The cicada has spent all summer singing, as they are wont to do, and is realizing he has nothing to eat. The ant on the other hand, widely recognized as hard-working providers of their nests with wisdom and forethought, has food aplenty because of his hard work. The cicada asks the ant (or ants in some versions) for some of his food. The ant declines his request and chastises the cicada for not storing up any food for the winter and, quite heartlessly, tells him that if he was foolish enough to waste the whole summer singing he can dance his way hungry through the winter. Ants possess many admirable qualities, but apparently compassion is not one of them. The cicada is a symbol of the costs of frivolity and laid-back living, whereas the ant is a clear symbol of the merits of industry and planning. The message of the worth of hard work and preparedness is clear.

MONSTERS AND OTHER NEGATIVE ASSOCIATIONS

The most ancient examples of feared and worshipped entities are nature-based or nature-inspired monsters. The definition of the monster as a mixed form is preserved in human-animal amalgamations that trace their roots to the animal cults of the ancients across varied cultures (Braudy 2016). Insect monsters are found worldwide in ancient folklore and mythology (Cherry 2012) and the insects serve as the basis of these monsters in part because of their otherworldly properties and nasty appearance or characteristics (they bite, sting, or are otherwise strange). Like the centaurs and satyrs of Greek mythology and the bird-woman valkyries and harpies, insect-based monsters often merge the strange and frightening aspects of insect attributes with the familiar traits of our own species to generate mildly relatable, yet frightening creatures of myth and legend that symbolize fear, evil, and dread in our stories.

Insects and related arthropods that inflict pain are generally employed to symbolize negative feelings of discomfort and dread or act as symbolic agents of punishment (Cherry 2002). For example, scorpions are mentioned ten times in the Bible, primarily as agents of pain and torment (Bruce 1992). These venomous arthropods generally represent evil and treachery in Israelite and Christian symbolism. For example, the Israelite King Rehobam, son and successor of King Solomon, in defiance of sound advice to the contrary, decided to increase the oppression on his people. He remarked:

> Whereas my father laid upon you a heavy yoke, I will add to your yoke. My father chastised you with whips, but I will chastise you with scorpions.
>
> (I Kings 12:11, Revised Standard Version)

This reference to scorpions is likely an allusion to another whipping instrument, perhaps one bearing metal barbs, that stings more than a regular whip. Some biblical commentary (e.g., Pfeiffer and Harrison 1962) describes a "scorpion whip" as one with barbed points or pieces of steel embedded in it and used for the scourging of slaves. This secondary association with the punishment of slaves would have been particularly offensive to the Israelite subjects. Rehobam's oppressive rule ultimately leads to the division of the Israelite kingdom into separate northern and southern kingdoms, with Rehobam retaining rule in the southern kingdom of Judah. On the other hand, this term may simply be an allegorical reference implying crueler punishment than they are used to at the hands of the previous king, King Solomon. Regardless, the symbolism is clear to anyone familiar with real scorpions.

People in the countryside east of the Mediterranean Sea must have had significant negative experience with these organisms for them to carry such recognizable disdain. Scorpions similarly figure, but in a more sensational way, as frightful agents of fear and obstruction in much older stories of this region and farther east into Mesopotamia (Black and Green 1992; but see Van Buren 1937–1939 for examples of beneficent scorpion symbolism in this region).

In the ancient tale of the Epic of Gilgamesh, the hero (King Gilgamesh) embarks on a treacherous journey filled with gods, goddesses, monsters, and other mythic creatures. Written in the Akkadian language in cuneiform script upon a series of baked clay tablets, this is arguably mankind's oldest known story. Apparently modeled after the historical King Gilgamesh who reigned in the Mesopotamian city of Uruk in about 2750 BCE (Mitchell 2004), the story's hero, Gilgamesh, befriends a naked, wild man (Enkidu) that enters into the world of the civilized through the erotic arts of the priestess Shamhat. After some antagonism, the cultured man of the city, Gilgamesh, and Enkidu, the symbol of wild nature, become good friends and traveling companions on a series of adventures. When Enkidu gets sick and dies, Gilgamesh is broken-hearted and in the anguish of facing his own mortality sets out on a desperate journey to find and consult the one man that can tell him how to escape the grip of death. This man is Uta-Napishti, the legendary primeval king of Shuruppak, a city on the banks of the Euphrates River; a man who survived the Deluge and because of his obedience was made immortal by the gods and granted a place among them (George 2003). Uta-Napishti and the story of the Deluge that destroyed the world are the precedent counterparts of Noah and the flood depicted in the Bible. To find Uta-Napishti, Gilgamesh travels eastward, toward the end of the world and the garden of the gods, where he resides. Nearing the end of his journey he must first pass through the mountains where the sun sets and rises, the mountains of Mashu. The way to the other side, a tunnel, is guarded by a pair of scorpion men (Figure 6.1). These were terrible and dreadful creatures whose glance was death and whose radiance was fearful (George 2003). What else would you expect of creatures that are part man (or woman) that also bear the metasomal "tail" and sting of a scorpion? Recognizing Gilgamesh as semi-divine, he is after all the son of the goddess Ninsun (his father was the priest king of Uruk, Lugalbanda), the guardians of the tunnel question him and decide not to kill him. Unable to convince him of the dangers that await, they give him instructions and allow him to pass. He runs through the tunnel, for twelve straight hours in total darkness, and ultimately arrives on the other side and makes his way to the garden of the gods. Uta-Napishti tells him the secret to immortality but acquiring this involves some difficult tasks. Deficiencies and mistakes on the part of Gilgamish result in his failure to complete the tasks and he does not become immortal. But he does succeed in rebuilding the semblance of civilization in the state of human affairs following complete destruction of the world by the Deluge.

Scorpion men also figure in the Babylonian creation myth (the *Enûma Eliš*) (Lambert 2013), being one of the eleven monsters created by the god Tiamat to revenge her husband's death. The morphology and role of this monster (called Girtablulu) is based on a different literary tradition than that described in the Epic of Gilgamesh.

In Mesoamerican antiquity, scorpions and their version of scorpion men played a similar, yet independently derived role. These people recognized

FIGURE 6.1 The hero Gilgamesh at the tunnel gate guarded by pair of scorpion men. Original drawing by Nicholas C. Williams, with permission.

scorpion constellations in the night sky that related celestial occurrences to their ceremonial customs. Scorpion behavior, particularly observed interactions associated with predation and courtship, was mirrored in the stars and evoked a symbolic connection to Mayan hunting, warfare, and sacrifice practices (Carlson and Cherry 1996). On a similar note, the scorpion-based constellation "Scorpius" was recognized by peoples of the ancient Near East and its status and related mythology incorporate known behavioral traits (e.g., dwelling in darkness, painful sting) of this arachnid (Carlson and Cherry 1996). People see their place and culture reflected (and perhaps justified) by what they see in the heavens.

The contribution of the scorpion form in all of these fictitious creatures symbolically represents dread and danger for those unfortunate to encounter them and are thus creatures to be avoided. Although scorpions were clearly the most feared arthropods of antiquity (Beavis 1988), other arthropods similarly carry negative connotations. For example, because of their association with putrefaction, death, and decay, flies have long been symbolic of these processes and associated negative phenomena. An ancient, familiar example illustrates this case of negative symbolism.

Baal was originally a supreme deity of the Canaanites, in the lands on the eastern shores of the Mediterranean Sea prior to the arrival of the Israelites. The name Ba'al is a common Semitic noun that means lord or owner. Over time, recognition of Baal spread to neighboring regions (e.g., Phoencia, Egypt) and became part of the regional pantheon. The worship of Baal became contentious in the eyes of the Israelites as a competitor of Yahweh (see II Kings 1), with the latter ultimately reigning supreme while at the same time subsuming characteristics of the former. Consequently, Baal is portrayed negatively in the Bible. The name Baal was combined with other words in the generation of new terms for places, deities, and ideas. One of these was the Philistine deity Baal-Zebub (or Beelzebub), worshipped primarily in the city of Ekron. Zebub is the collective term for flies and hence the basis for interpreting Baal-Zebub as the "Lord of the Flies." Adherents of this interpretation consider this god to have dominion over flies with the power to both cause and cure disease and to be able to foretell the course of a particular disease or ailment in a person (Cloudsley-Thompson 1976). The ancient Mesopotamian god of the underworld, disease, and death was Nergal, one symbol for which is the fly (Black and Green 1992).

On the other hand, good etymological and contextual arguments exist for variant spellings and therefore variant meanings (e.g., lord of the divine abode, Baal the prince) of this moniker that have nothing to do with flies (van der Toorn et al. 1999). In this case, biblical references may be a purposeful distortion of the term in order to demonize this alien deity by symbolically saddling Baal with the negative connotations of flies. Later (see New Testament references), Baal-Zebub became synonymous with Satan.

Our innate, negative impressions of insects have been exploited to symbolically convey dread, disgust, and other unsavory feelings. It is not surprising then that insect symbolism has been employed in the human struggle with that most uncomfortable of events, our inevitable death.

DEATH

Humans have a long and tortuous preoccupation with dying. Death is inescapable and yet we wish to avoid or conquer death. Strong religious beliefs are apparently associated with a fear of death (Ellis et al. 2013) and the origins of some of these belief systems may stem from their ability to help ease human anxieties associated with dying. Because death and what happens when one dies is inherently associated with the unknown, it is not surprising that human concepts of death and a hoped-for afterlife are something difficult to convey and thus are filled with analogy and symbolism. The small size and foreign morphology of insects lends them to be hard to relate to and thus relatively unknown and misunderstood and generally feared. Therefore, insects are good candidates to symbolize ignorance-based fears in people. Consequently, insects play no small role in the symbolism of death and human notions of life after death (Cherry 2011).

One group whose culture was influenced by a strong preoccupation with death and the idea of an afterlife was the ancient Egyptians. They lived in northeastern Africa primarily along the lower Nile River and its delta at the Mediterranean Sea. Their civilization flourished from around 3000 BCE until the end of the Ptolemaic dynasty with the death of Cleopatra VII in 30 BCE. For thousands of years this culture developed a very sophisticated and complex belief system comprising a myriad of deities in their attempts to find meaning in their existence. Their understanding of death and resurrection was strongly influenced by the biology of a local insect. This insect took on great symbolic meaning for these and nearby peoples and became what is arguably the most well-known and iconic insect-based symbol of antiquity—scarab beetles.

In the landscape familiar to the ancient Egyptians, there were several genera of dung beetles, members of the family Scarabaeidae, subfamily Scarbaeinae, that as adults collect freshly deposited feces of large mammals. During this time, the dung was primarily generated by cattle, a supremely sacred animal dating back to the earliest periods of Egypt's history. Female bovine deities represented the powers of creation and fecundity (Wilkinson 2003). The beetles use the dung to feed themselves and their young. Three local genera (*Kheper*, *Scarabaeus*, and *Gymnopleurus*) are dung-ball makers and rollers whose activities were well known to the observant Egyptians (Ratcliffe 1990). Most notably among these is the Sacred Scarab (*Scarabaeus sacer*) (Figure 6.2a), a species found in the area's coastal dunes and marshes. The morphology of this beetle is notable here in that the leading edge of the adult is somewhat evocative of a stylized sun. Projections on the side of the face and clypeus, and the fore-tibial teath when the front legs are held at rest, resemble radiating beams of light from the sun (Figure 6.2a). Singly, or more typically in nuptial pairs, these beetles move dung away from the activity and competition of the acquisition site by forming some of the dung into a sphere and rolling it away. It is then buried in a chamber some distance from the soil surface. The dung is either fed upon by the adult beetles or it can be used to provide for the next generation. If the latter, the beetles mate and an egg is laid on a pre-formed brood ball. The larva

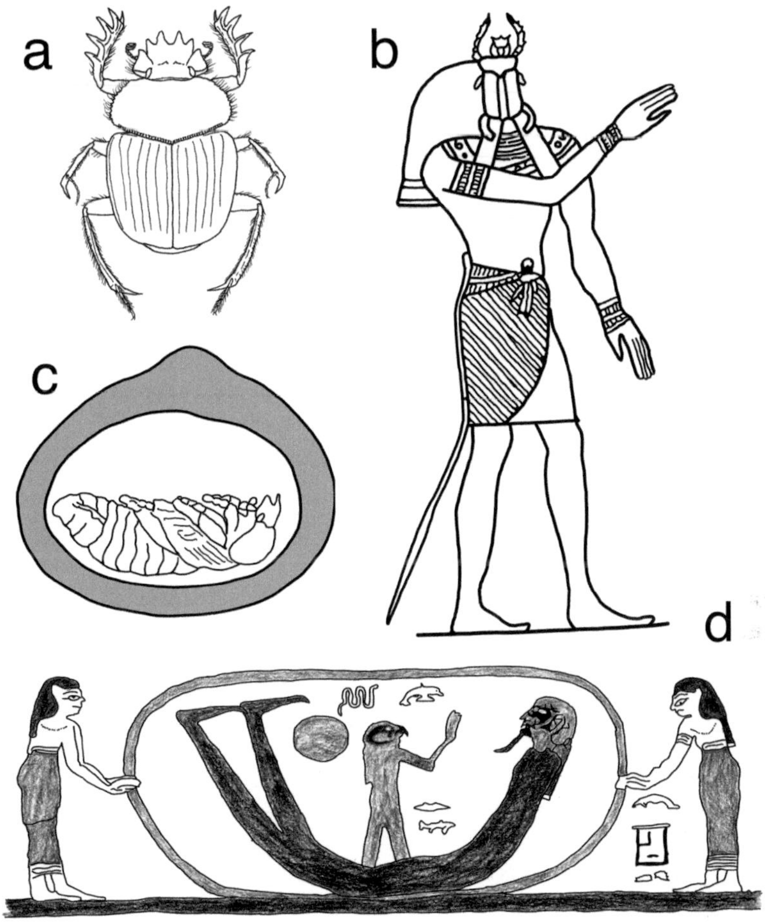

FIGURE 6.2 Dung scarab cult of ancient Egypt. a) *Scarabaeus sacer* (Coleoptera: Scarabaeidae); b) sun god Kheper; c) pupa of dung beetle in chamber of brood ball; d) Osiris in repose attended by son Horus (center), wife and sister Isis (left), and Nephtys (right), sister of Isis, from tomb of Ramses VI. Drawings based on various sources. Redrawn by James N. Hogue.

feeds on the dung provided, grows, and ultimately pupates in the underground chamber. Finally, an adult beetle emerges from the ground. Their morphology and biology therefore unite three important elements: the sun, the soil (earth), and cattle. This beetle was naturally predisposed to serve as a focal symbol of ancient Egyptian sun-cult theology.

The sun god Re (a.k.a. Ra) is among the most important deities, granted with much responsibility and power and whose activity occurs within the heavens, the earth, and the underworld. Re directs the three seasons of the Egyptian year: inundation of the land by the Nile River, the growing season, and harvest season. Re also created other deities and human beings and is considered the father and mother of all living things. This god takes on different forms and has over time merged qualities and appearance with other solar and cosmic gods. The sun god takes on three forms: Khepri in the morning, Re at midday, and Atum in the evening. Khepri is thus the god of the rising sun, a sun that is perpetually recreated at the dawn of each day. Khepri is thus associated with the concept of resurrection. Khepri is often depicted in the semi-anthropomorphic form of a man with a scarab for a head (Figure 6.2b). Just as dung-rolling scarabs roll spheres across the landscape, Khepri moves the sun from darkness to light, up into and across the sky. Furthermore, new, living beetles arise from the earth much as the resurrected sun arises from the edge of the earth in the east. These people observed adult beetles bury dung. Later, the beetles died but were resurrected from the ground. From death comes life. Surely these creatures must hold the secret of life. The beetle, in life and in symbolic form, represents this regenerative cycle and this process is realized in the beetle. For example, the scarab hieroglyph represents the verb kheper that means "to create" or "to become" (Southwood 1977). Analogous traits beget affinities that then become symbolically interchangeable (Cashford 2003).

An additional association is between the mummy, the god Osiris, and the dung beetle pupa. Osiris is a god of the earth and a god of the dead. He is also a god of death, resurrection, and fertility. He is of human origin but was murdered and brought back to life. Over time he took on attributes of other gods including his role as a funerary deity. He is closely associated with the sun god Re, functioning as the counterpart of Re in the netherworld. It is as if Re is the earthly body and Osiris is the soul, re-uniting with Re during the night-time part of the solar cycle. Whereas Re is a lord of the heavens, Osiris is lord of the underworld (Wilkinson 2003); a resident of the heavens during the day and of the earthly realm at night. This exemplifies the recurring theme of the cycle of death in the earth and life in the heavens. When combined as Re-Osiris, this deity is sometimes depicted as a man with the head of a beetle. Osiris is also the father of Horus, a product of his mating (after he had been resurrected from the dead) with his sister Isis. Horus is the focus of some of the most extensive mythology of ancient Egypt and is visualized in many forms but typically as a sun god, again a god of the rising and setting sun. Horus is lord of the sky and is regularly depicted in falcon or leonine form, often as a falcon-headed man.

The exarate beetle pupa (Figure 6.2c) in the brood chamber is the model for Osiris (Figure 6.2d). It lies motionless and dead-like, its developing legs lay exposed and held folded and nearly crossed over the chest in like manner of a human mummy. Humans must also go through this process of life, lifeless pupal or "mummy" stage, and rebirth in imitation of what was understood of the scarab life cycle and of the solar cycle of death and darkness followed by solar resurrection (Cambefort 1987). The religious framework of the ancient Egyptians was organized around this idea: birth (egg), life (larva), death (pupa), and resurrection (adult) stages in the life of a person that mirror the stages of existence of the beetle (Harpaz 1973).

This symbolism linking insect biology with burial practices and hoped-for resurrection of the dead from a mummified corpse comes to full fruition when one considers the pyramids containing the tunnels and tombs of the royalty wrapped up as mummies (scarab pupae). This led some workers (e.g., Cambefort 1987; Hanski 1992) to symbolically equate pyramids with venerable masses of dung. The tunnels, passages, and rooms of the tombs are like the underground tunnels and brood chambers of the beetles. The site of new life (resurrection) is the pyramid for humans, as it is the chambered brood ball for the beetles.

Carved scarabs are the most popular styles for Egyptian amulets. Some were placed on the mummy of the deceased whereas others adorned the tomb and sarcophagi or were the models for temple sculptures (Wilkinson 2003). Over time, scarab imagery went from funerary objects and paintings adorning the tombs of the dead to a variety of amulets, seals on finger rings, bracelets, and pectorals (broad, highly decorated elements hung from the neck over the breast) worn by people (e.g., Vilímková 1969). Carved scarabs often bore inscriptions on their ventral surface such as "good health," "life," and "my heart is true" (Klausnitzer 1981). The morphological detail on many carved scarabs is such that the model genus can be determined. The scarab image was even portrayed by the ancient Egyptians among the stars in what we now refer to as the constellation Cancer (Staal 1988).

As was true of later Hellenistic and Roman influences, Egyptian religious ideology spread beyond the borders of Egypt and was incorporated into the religious thinking and practices of the greater region. Knowledge and religious incorporation of ancient Egyptian sun cult theology existed among the inhabitants of the lands east of the Mediterranean Sea, including the Israelite people. This was facilitated in no small degree by the proximity to Egypt, the past residence of this people in Egypt, and the local presence of the same and related beetles (e.g., Rittner 2011). For example, a good argument can be made that a well-known vision experienced by Ezekiel was inspired by his familiarity with dung beetle morphology, behavior, and associated theology (Hogue 1983). Early in the sixth century BCE, around the time of the destruction of Jerusalem and imposed exile of many of its inhabitants by the Babylonians, Ezekial (see the Old Testament book Ezekiel, Chapter 1) describes a vision that is meant to convey the glory of God and his exaltation above creation as part of a longer tale of doom and forewarning to the Israelite people about to realize their impending fate. He carefully describes a highly symbolized representation of a higher order angelic figure, perhaps a cherub, and an associated set of wheels. His descriptions of the angel's face, legs and feet, wings, surface characteristics, and behavior are consistent with a sometimes-stylized depiction of a dung scarab. The wheels within wheels, capable of omni-directional mobility, symbolize God's omnipresence and influence and could very well be inspired by a layered sphere of dung pushed around by the beetle. Ezekial was a learned person, a priest and a prophet. He would have been intimately familiar with the theophany of dung scarabs and the Egyptian pantheon and thus easily inspired by the symbolic imagery of these deified beetles.

Lest we feel tempted to assign all of this talk of beetles and celestial relationships to unsubstantiated myth it is worth noting that these insects do have a real connection to the heavens. Ball-rolling dung beetles perceive the stars in our galaxy and use the band of light across the night sky we refer to as the Milky Way as a stellar landmark by which they navigate their path across the landscape when rolling their balls of dung (Dacke et al. 2013).

In China, cicadas also serve as symbols of resurrection, and for much the same reason. Adults of these insects were observed to come from, apparently generated by, the soil of the earth. Carved cicadas (tongue amulets) were put on the mouth of a dead person to aid them in obtaining new life and immortality (Riegel 1994; Pemberton 2003). Cicadas are modeled in or their images adorn ancient Chinese bronze vessels, pottery, jade sculptures, and other objects in testament to their symbolic importance to these people at this time (Riegel 1994). An identical symbolic role for cicadas, apparently based on the same connection with emerging alive from the earth, is found in ancient Greece (Egan 1994) and among some native American peoples (Clausen 1954). The emergence of the larva from the soil of the earth and the shedding of the larval exuvium, followed by the development of the adult body and the expansion and subsequent use of wings, was described by esteemed philosopher of ancient Greece, Aristotle. Later Greek writers elaborated on cicada life history, including their apparent return from the world of the dead, and the symbolism of immortality. Such observations of arising new from the earth invoked a sense of rebirth and resurrection as perhaps for a buried body. This association and symbolic representation involving the cicada life cycle were very influential in the ancient Greek's beliefs about birth, life, death, and rebirth (Egan 1994).

This symbolism also finds its way into the epic stories of the Greeks, such as the story of Eos and Tithonus in the Homeric hymn to Aphrodite. Tithonus, as depicted in one of the Homeric hymns to Aphrodite, is the son of King Laomedon and thus the prince of Troy. He was the lover of Eos (Lady Dawn, Goddess of the Dawn) who appeals to Zeus to make her young lover immortal. The supreme god grants her wish, but Eos neglected to ask that he not grow old. Tithonus is thus immortal but is not given the gift of eternal youth. He grows old, and once too decrepit to move is shut away in a room by Eos (Godolphin 1964). In later versions, Tithonus withers away, his body decreasing in size until he ultimately turns into a cicada. Now he can live and sing forever (Egan 1994; Janko 2017). Being a cicada is thus equated with immortal life.

Because the New Testament was written by people significantly affected by Hellenistic culture, it is interesting that given the importance of cicadas in Greek mythology, folklore, and symbolism, especially with respect to resurrection and a new life after death, cicadas are not mentioned or used in this fashion in the Bible.

THE SOUL

Associated with most understandings of death, dying, and travel to another realm of existence is the idea of the soul. Envisioned as a spiritual or immaterial manifestation of one's self that leaves the body at death, the notion of a soul is widespread across many cultures. It can embody an eternal existence or serve as the vehicle transporting one's essence to another world in the afterlife. The soul itself has long been depicted as a flying insect of various sorts including bees (Cook 1895), butterflies and moths (Beavis 1988; Cherry 2011), cicadas (Egan 1994), and flies (Cherry 2011). Symbols of the soul in ancient Greece are butterflies, but more often souls are represented by moths (Cherry 2011). In the ancient Egyptian Book of the Dead from the Eighteenth Dynasty (1555–1350 BCE) the praying mantis serves as the deified escort of human souls to the netherworld of divine spirits (Prete and Wolfe 1992).

That adult Lepidoptera provided a means for the soul's ascension to the next world was almost certainly inspired by the silent and ethereal nature of flight in some butterflies and moths, especially those active at night. Moths are well known to be attracted to lights, much like the souls of recently dead that travel at night looking for light (Clausen 1954). This nocturnal motif is reinforced by the widespread notion of the moon being an image of the soul and as an abode for the soul after the body dies (Cashford 2003). The Greek word Psyche came to mean both soul and butterfly; and the goddess Psyche, who represented the soul, is typically depicted with lepidopteran wings (Kritsky and Cherry 2000). Cicadas also represent human souls in ancient Greek beliefs, again likely founded on observations of winged creatures emerging from the earth, thus going from one realm of existence to a higher existence as a winged creature that can ascend to the heavens (Egan 1994).

Whereas the ancient Egyptians did their best to keep flies away from dead bodies before they could be mummified, other cultures encouraged the activities of flies. The Moche, a pre-Incan people that lived near the northern coast of Peru during the first through seventh centuries CE, purposely exposed their dead to flies. The maggots consumed the body and spirit of the deceased. Once transformed into flying adult flies, they carry the soul of the deceased and their imminent close contact with living humans completed the cycle of life (Huchet and Greenberg 2010). Images of flies on ceramic vessel art served as symbols of this insect's role in this culture's understanding of death. Scenes of dancing skeletons being watched over by a fly and of flies watching over a parade of chained, naked, and soon to be dead prisoners (see Huchet and Greenberg 2010) clearly establish the link between flies and what becomes the dead.

Some Indigenous peoples in central Russia, in the area of the Ob River, practiced shamanism and held religious beliefs incorporating a close affinity to nature and respect for ancestors. Among these are the Khanty and Mansi peoples. Here each person consists of a physical body and several souls. Their notion of the soul involved a somewhat sequential series of souls (five for men and four for women) following death that continue to interact with the living (Jordan 2003). Following the shadow soul (also known as the grave or buried soul), which remains with the decaying body until it is completely decomposed, the soul is transformed into a beetle (Chernetsov 1963). This idea may have its roots in the appearance of beetles during the latter stages of body decomposition and encountered around graves sometime after burial. In some locations, the kind of beetle is indicated and two of these are aquatic beetles (hydrophilids and dytiscids). In this context, these beetles thus symbolically represent a vehicle for travel of the soul to the next state of existence, and this is via water. Water is nearly universally associated with the source of creation and among these peoples water also possesses a soul (Jobes 1962) and is the route to the underworld. This status, being active souls of once-living people, offered these beetles protection and it was not good to harm or kill them. They apparently were also useful as soothsayers that were consulted by young women to find out when they would marry (Jobes 1962). Interestingly, the next soul (the bird or sickness soul) is thought of as a person or bird, and sometimes as a mosquito. This soul may wander or travel to the underworld, from which it may or may not return and it may also ultimately be transformed into a beetle (Jordan 2003).

RELIGIOUS SYMBOLISM

Ancient peoples considered honey bees and their products as sacred (Crane 1999) and their characteristics worthy of emulation. In religions that required offerings as acts of devotion, honey was a common item sacrificed to the gods. This was true of early honey-hunting peoples that got honey and beeswax from naturally occurring hives. After people began keeping bees in managed hives, and thus became more familiar with the biology of these insects, bees took on an important symbolic role (Crane 1999). This happened among people of ancient Egypt, Greece, and Mesopotamia and was incorporated among people practicing long-held indigenous religious beliefs and more derived theologies such as Judaism, Christianity, and Islam. Related eusocial bees (Apidae, Meliponinae) attained similar status in the New World among such people as the Maya. Symbolism incorporating these bees and their products served both illiterate and literate people in their worship and the understanding of associated abstract concepts. Leaders during the early centuries of the Christian church wrote in praise of honey bee colonies and highlighted the apparent orderliness of their life and work and thought of this as a worthy model for a Christian community. Dating back to the works of Aristotle (1910), there may have existed a bit of a misunderstanding regarding the sexual life of bees and what bee, or bees, were the leaders of the hive (Davies and Kathirithamby 1986). The colony was thought to be led by a male bee (the king) and because nobody had seen bees having sex, it made sense that bees were the result of asexual reproduction via virgin birth. The jump to a symbolic association with human nobility and leadership, and with the virtues of chastity was short and likely inevitable.

Prominent early religions of western Asia and eastern Europe, especially under the Hittite influence during the Bronze Age, worshipped a great goddess of life, death, and regeneration (Great Mother Goddess, Fertility Goddess). She was thought of as youthful and beautiful and was associated with fertility and motherhood. She was omnipresent in the natural world and reflected the purity of untouched nature (Gimbutas 1974). This goddess is well represented in figurine art throughout the Neolithic and Chalcolithic eras in this area (see examples in Gimbutas 1974) and symbolically via depictions of bees (or sometimes butterflies) (Figure 6.3). This symbolism is best developed in Minoan and Mycenean art from the early to mid-second millennium BCE and later (see also Evans 1925).

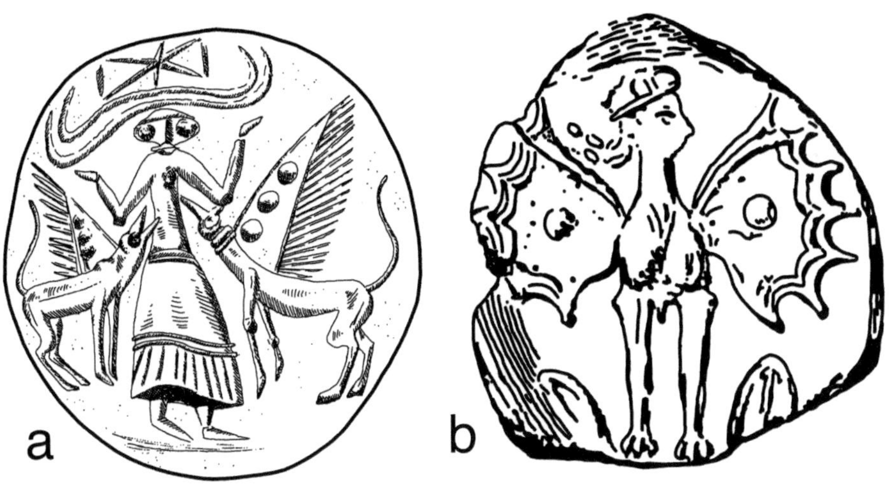

FIGURE 6.3 Bee and butterfly seals. a) Bee goddess on engraved onyx stone (Knossos, Crete ~1500 BCE), redrawn from Gimbutas (1974); b) butterfly seal depicting eyed butterfly wings on body of sphinx (Zakros, Crete). Redrawn by James N. Hogue from Evans (1925).

Cattle, particularly bulls, have a long and widespread tenure as sacred animals. They are symbolized and associated with many phenomena, including the fecundity of nature (Jobes 1962). Partly because of their horns resembling the crescent moon they have a symbolic lunar association (Gimbutas 1974; Cashford 2003) and in this context they symbolize cycles of periodic regeneration and fertility of the Great Goddess. Bees, partly because their antennae resemble bull horns, their periodic swarming, their creative activity of producing honey (the food of the gods), and probably more so because they were thought to be spontaneously generated from the rotting carcasses of bulls or oxen (Ransome 2004; see Osten Sacken 1893 for interpretation) and thus had their origin in these sacred animals, are why manifestations and representations of this goddess of regeneration and fertility took on the likeness of the honey bee.

The Great Goddess existed and was worshipped for at least five thousand years and was referred to by a variety of names before being incorporated into classical Greek civilization. Under Greek and Roman rule, she became known as Artemis. This is not the truly Greek Artemis, twin sister of Apollo, the counterpart of the Roman goddess Diana, but an integration of the archaic cult figure with a Greek deity. This became an Ephesian cult with a strong Greek character that was widely known. The Temple of Artemis is referenced in the Bible (Acts 19) and was considered one of the seven wonders of the ancient world. This institution even took on a symbolic organization analogous to a bee hive with swarms of virgin priestesses (*Melissae*) and male officials (perhaps eunuchs) that were called Essenes (meaning "King Bee") who practiced chastity during their reign (Gimbutas 1974; Ransome 2004). Again, some of the discrepancies between this idealized representation and reality is because of an early misunderstanding of the sexual life of bees. Despite such misconceptions it is worth noting that humans recognized all the stages of insect metamorphosis and generally incorporated this into their symbolic constructs from at least seven or eight thousand years ago (Gimbutas 1974).

Even more so than the Greeks, the people of ancient Egypt have a long association with honey bees and beekeeping. Although people have hunted honey for at least eight thousand years, it was the Egyptians of nearly five thousand years ago that first practiced large-scale, sophisticated beekeeping (Crane 1999; Kritsky 2015). This enterprise was important to their lives and honey bees were revered for their qualities and products. This in turn resulted in the imagery of bees being incorporated as symbolic representations of different aspects of the Egyptian cultural experience. For example, the sun god Re cried and his tears, falling to the ground, turned into bees. These insects then went about their work, visiting flowers and producing honey and wax. Bees therefore have a divine origin.

Early on, ancient Egypt was divided into two regions along the north-flowing Nile River: Upper Egypt to the south and Lower Egypt to the north. Perhaps because beekeeping was most important in the delta region (Crane 1999), the bee became a symbol of this region and thus of Lower Egypt. A sedge plant (Cyperaceae, perhaps *Cyperus rotundus* or the domesticated *Cyperus esculentus*) (Crawford 2007) became the symbol of Upper Egypt. The two regions of Egypt were united around 3100 BCE and the king was now a dual king of both Upper and Lower Egypt, and thus literally the king of sedge and bee.

REVERENCE, ASSOCIATIVE, AND BESTOWMENT SYMBOLISM

People use insect symbols as a means of touting various, often virtuous, characteristics of the insects themselves. Predictably, people then use these symbols to associate these attributes with a person or people or as a means to hopefully bestow that trait on them in the near future. The model possesses some trait that is desirable to a person, such as beauty or strength. Employing a form of imitative magic (Cherry 2005), association with the model, directly or via a symbolically charged representation, serves as a means of character bestowment from the model to the human participant. Several examples follow that illustrate such associations.

Insects have been used to exhibit both positive (industry, beauty, persistence, power) and negative (decay, laziness, fear) ideals or feelings. One place that insects are depicted as symbols of traits in antiquity is on coinage. For example, animals associated with various deities of the Greek pantheon are portrayed, sometimes symbolically, on Greek coins. These symbols often came to serve as badges that represent a certain city and additionally served to indicate the origin of a particular coin. For example, Artemis, the patroness of the city of Ephesus, is symbolized by the bee on coinage of Ephesus and is clearly indicative of its origin from this city (Ransome 2004). In other cases, animals are found here as representations for more aesthetic or idealized purposes. This is typically the case of most insects on these coins. On coins from the ancient Greek colony of Metapontum, in what is now on the southern shore of Italy, are found images of grasshoppers, ants, mantids, and cicadas (Fagerlie 1964).

Head hunting, primarily in ongoing cycles of revenge with neighboring groups and as manhood rights-of-passage rituals, has a long history among many cultures in the South Pacific, including the Asmat people of southwestern New Guinea. These people were familiar with mantids as large predators that voraciously consume their prey, often while it is still alive. They may also have been aware of the sometimes-observed behavior of praying mantid females consuming at least part (invariably the head end) of her male counterpart during or following copulation. Cannibalism was intertwined with head hunting, so it is not surprising that praying mantids served as symbols that objectify the complex decorum of Asmat head-hunting practices. For example, carved mantids adorn the figureheads at the prow of some war canoes (Cherry 2004).

In like-minded fashion, insects that symbolize other sought-after traits can bestow such traits to a person by symbolic association. For example, the transference of the ferocity of wasps to warriors in ancient Greece was mediated by symbolic imagery (Davies and Kathirithamby 1986). Other desired traits of these insects include their homebuilding skills.

In Mesopotamia and ancient Egypt, the fly stood for impudence and for courage. Fly amulets and pendants appear frequently in these areas in ancient times (Harpaz 1973). In the New Kingdom period of ancient Egypt a mark of rank, the Order of the Golden Fly, was bestowed as a military honor on some people. The Egyptian Queen Ahhotep, herself a fierce war commander, was awarded these golden flies of valor on a necklace, a necklace that later adorned her mummy. Some fly pendants were found as treasure buried with royalty and their wives (e.g., of Tuthmosis III) (for examples see Aldred 1971). The fierceness and persistence of some flies (e.g., *Stomoxys calcitrans*) were qualities revered by people and such awards associated these traits with the bearer.

There is a widespread vision of the moon representing the cycle of life. Everyone on the planet is familiar with the moon and knows that every month it comes and goes, only to be regenerated with the next lunar cycle. It is a symbol of the cycle of life and fertility. The moon spins, with its invisible threads, a web of relationship between all created beings of the earth which are similarly bound to the cyclical rhythm of waxing and waning, the law of the order of life (Cashford 2003). It spins the proverbial web of life. It is not surprising then that the spider is used as a symbol of the moon and its spinning activity in many cultures including the ancients of Egypt, Greece, and Europe as well as Indigenous peoples of western North America, Central America, central Brazil, Indonesia, the South Pacific, and southwestern Africa (Cashford 2003).

People of the Blackfeet/Siksika nation of the northern Great Plains of western North America sometimes incorporated a symbol representing a butterfly on their tipis (Figure 6.4). This symbol resembles a Maltese cross with apically truncate arms and is used to bring sleep or dreams to the inhabitant of the tipi (Grinnell 1899).

The tremendous powers of reproduction are seen as favorable characteristics in some insects, and they have come to symbolize fertility and fecundity. Association with, including consumption of, these insects is thought to bestow these qualities upon a person. This is the case for bees (Ransome 2004) and termites (Fairhead and Leach 2003).

Totemism is an extension of this type of symbolism. Totemic representation of insects may symbolize ancestry or kinship. This generally leads to feelings of adoration of or reverence for the organism depicted and these models are

FIGURE 6.4 Blackfoot/Siksika Tipi and butterfly and symbol. Tipi drawn from photograph by Walter McClintock, late 1890s in Montana, housed in the Yale University Library. Inset butterfly symbol redrawn by James N. Hogue from Grinnell (1899).

often associated with ancient stories and myths integral to the cultural history of a group of people (Berenbaum 1995; Cherry 2002). Totemic incorporation of insect symbols is well represented by Indigenous cultures of northwestern North America (Kritsky and Cherry 2000) and Australia (Bodenheimer 1951).

Insect symbolism is sometimes indirectly inspired or linked to the insects themselves. This is the case of fly whisks. Originally developed as a tool to shoo flies, these usually short-handled tools are fitted with long hairs of animals (often the tails of large mammals) or plant fibers. Over time these practical utensils evolved a less utilitarian function as a symbol of authority in some cultures (e.g., both sub-Saharan and northern Africa, Pacific Islands) (Riegel 1979). The handles were sometimes adorned with decorative objects, precious materials, and symbolic images of animals including in one case, and most interestingly, with a spider. The tool, designed to get rid of insects, took on the role as a symbol of social status. Another example of indirect association in insect symbolism involves the use of insect products in the production of highly symbolized objects. Sometimes beeswax was used as the material from which a symbolically charged object was sculpted. In other cases, such figurines were cast in metal using the lost wax technique where the image was first modeled in beeswax. This model then was used to create a mold that, after the wax was removed by melting, a negative-image space was created that was subsequently replaced by molten metal. Beeswax was employed in this way by ancient cultures nearly worldwide (Crane 1999).

CONCLUSION

Because of their apparentness, strange yet curious familiarity, and influence on pre-historic people, insects and related terrestrial arthropods have found widespread use in the symbolic communication of ideas in human cultures of antiquity. Modern examination of this use of symbolism provides insight into the development of the human mind and subsequent cultural ideals surrounding a people's understanding of self and their place in the cosmos. Insect symbolism is sometimes used to facilitate positive or utilitarian human affairs such as love, commerce, or politics. More commonly though, it involves more mysterious things, particularly in our distant past. From the dawn of human culture, insect symbolism is part of how we came to develop and disseminate abstract notions related to an understanding of things like death, disease, tragedy, and other fearful or otherwise mysterious phenomena previously more dreadful when in the grip of realized ignorance. These are the roots of our monsters. Modern man is cognizant of this and knows better, yet still seems unable to shake the fetters of what got us here. Symbolism, including insect-based symbolism, is still used to manage and convey a variety of ideas and our penchant for monsters that keeps alive our fascination with the primitive, the unfamiliar, and our comfortable fear of things unknown is alive and well. As is true of Sasquatch and the Loch Ness Monster, the days of insect symbolism and insect-based monsters have yet to wane.

CHAPTER SEVEN

Insects in Literature and Language

Literary Insects of Greek and Roman Antiquity

RORY EGAN

Greek and Latin texts surviving from the millennium extending from the beginnings of Greek literacy to the onset of the Latin Middle Ages are infested by insects with a diversity and frequency that probably exceed what is typical of literary history in general. Even veteran readers of classical literature might be surprised by this, their attention being dominated by mythical exploits of gods and heroes, the protagonists of epic and drama, by pioneering eminences of intellectual history, or by the amatory intrigues and politico-military activities of kings, queens, emperors, and generals. Yet the combined lexical records of the two languages show several hundred names or terms for insects or other small creatures such as arachnids and annelids which, for literary purposes, can be accorded honorary status as insects. Some of the terms are redundant, others imprecise in their designation, and many have no noteworthy literary presence. It is, then, a significant minority that has gained literary or critical prominence in notices that range from brief allusions to leading roles as protagonists in comedy, mock-epic poetry, philosophical drama, moral essays, or rhetorical parody. Insects do haunt virtually every prose or verse genre and the writings of such eminences as Homer, Sappho, Aesop, Aeschylus, Aristophanes, Plato, Aristotle, Plautus, Lucretius, Vergil, Horace, Ovid, and Seneca. Factors affecting the degree of prominence achieved by a literary insect include the extent to which its real-life ethology, physical appearance, or life cycle impinged, pleasantly or unpleasantly, on the experience of writer and audience; hence social and/or "musical" insects receive the bulk of the attention. Other factors are the artistic ingenuity of an author and the perpetuating force of literary tradition, for Greek and Roman writers of all periods were encultured by the customs of their craft to emulate and transvalue their predecessors' creations.

The present format and the wealth of material in the centuries-long record dictate drastic culling and pruning. Regrettable though the omission or cursory treatment of many artful and interesting works or passages might be, it is a consolation to the indebted author of this chapter, as it must be to its readers, to have available two English-language resources on insects in the literary and visual arts of ancient Mediterranean culture (Davies and Kathirithamby 1986; Beavis 1988). One of them is restricted to Greek materials while the other recognizes the contiguities of Greek and Latin literature. Together they document the range and amplitude of the evidence in an indispensable encyclopedic platform for further study. The three decades since their publication have also seen scholarship on the insects progress in company with critical interpretations of the literary matrices themselves; sometimes in the form of well-documented studies of particular types of insects (Pataki 2013; Berrens 2018). Of that there will be intermittent notices in what follows. Given the obligatory selectivity, it is convenient to approach the corpus of ancient Greek and Latin literature as a complex organism within which themes and conceits circulate between eras, authors, genres, languages, and localities, and from one religious, philosophical, or scientific context to another. That affords both a convenience and a rationale for following a somewhat meandering course when any simple route dictated by literary categories, chronology, or entomological taxa would be impracticable. Much of the subject matter, after all, is timeless and, given the randomness with which the literature of the period survives, we can rarely be confident that the historical progress of narrative structures and figurative conceptions or adaptations matches the known chronological sequence of the literary documents.

Another delimiting factor is the necessary screening out, but for occasional token notices, of the centuries-long reception history running through Byzantine Greek and Medieval Latin literature, and the vernacular literatures of the Renaissance and modern times right up to the present. It must also be noted that the vastness of that post-classical literary tradition is counter-balanced by a lengthy pre-classical prodrome in which the ancient Mediterranean civilizations were partners in an extensive cultural continuum. This overarches the traditional demarcation of Orient from Occident and stretches back over another two millennia or so into Egypt and western, even central, Asia (West 1997). Our still-growing knowledge of cultural interrelationships between Mediterranean and Near Eastern civilizations encourages the notion that the oral traditions of the latter as well as the literatures that emerged from them will have exploited insect-related narratives and conceits.

Verification of such inter-cultural traffic in insect lore over space and time takes various forms. A small sampling would include the report, garbled or fabricated, relayed by the fifth-century historian Herodotus about furry ants in northern India, larger than foxes, who mined gold in the desert and ferociously guarded it. Of quite a different sort are a couple of similes (to which we recur later) from the *Iliad* that liken warriors at Troy to bees and flies which scholars have compared with passages from the Hebrew books of Deuteronomy and Psalms in one case and in the other from the much earlier Akkadian *Atrahasis* and *Gilgamesh Epic* (Davies and Kathirithamby 1986: 51, 151, n. 125). Mesopotamia also yields cuneiform texts dating from the early second millennium BCE that record moralizing animal fables of the Aesopic type. That form of parable, both entertaining and instructive, appears also in early phases of surviving Greek literature, specifically the diverse works of such seventh-century poets as Hesiod and Archilochus, earlier than any Greek prose text attributed to anyone named Aesop. The animal fable, indeed, with its large complement of insect protagonists, forms a continuum of its own that extends across the millennia from ancient Sumer right up to the present (Adrados 1999: 287–364). It has a conspicuous literary representation from classical times that includes the prose collection attributed to Aesop, but also such smaller selections as the poetic versions by Babrius (second? century CE) in Greek and Phaedrus (early first century CE) and Avianus (*c.* 400 CE) in Latin. Apart from such collections, there are of course numerous insect fables dispersed individually across the spectrum of Greek and Latin literature. Some among these are moralizing allusions rather than full narrative treatment. Examples of such are to be found in the biblical book of *Proverbs* (Davies and Kathirithamby 1986: 42, n. 4), a comparatively late instance of Near Eastern literature that was adopted into the Mediterranean cultural milieu through translation into Greek and subsequently Latin.

In one respect the early diffusion of animal fables might have had a more broadly pervasive effect on the representation of insects within the cultural continuum because, as in fables, the insects of other genres tend to be humanized in their actions and motives. One early instance of this is a poem, *sui generis*, written in iambic meter (typically used for poetic invective) by Semonides of Amorgos in the seventh century BCE. Semonides presents a catalog of animals each of which he matches to a type of woman with comparable behavioral traits. Semonides' correspondences are always pejorative except in the case of the bee-woman who is commended for qualities desirable in a wife. This is only one out of a wide range of exemplary literary roles assumed by the quasi-human bee from the time of Homer down to the end of antiquity.

The longevity of the humanized insect in fable evokes, if only accidentally, thoughts of Tithonus, a rare example of an insect with a personal name, who is long-lived in a double sense, given his place in western literature from Archaic Greece down to and beyond his role as titular persona of Tennyson's poem; given, too, the endless life granted him in literary myth. He also exemplifies the intricate variations that can affect an insect's course through cultural history. His name occurs first in the *Odyssey* and then a little later (seventh and sixth centuries) in several passages of poetry including, notably, some lyric lines of Sappho which have only recently been published after their recovery on papyrus that Egyptian sand had covered for two millennia or more (Janko 2017). The same name occurs sporadically in the literature of the next several centuries and eventually with great frequency in Latin poetry where it alludes metonymically to the east from which personified Dawn (Eos in Greek, Aurora in Latin) arises daily. The underlying story has Tithonus, an extraordinarily handsome and musically adept prince of the Trojan royal family, amorously captivating the Dawn goddess who spirits him away to the East, makes him her mate, and has immortality bestowed on him. His eternal life, though, is not packaged with immunity to aging, so he deteriorates in strength and physique as life drags on. Eventually, Dawn remedies the situation by arranging his transformation into a cicada. A key factor here is the coincidence that the Greek word *gēras* means both "old age" and "cast off skin." This meshes with the popular notion, that cicadas sloughed off old age to gain rejuvenation or immortality by shedding their integuments in the molting process (observed and described by ancients including Aristotle and, allusively, Plato).

Encultured familiarity with such traditional stories forged an implicit contract between audiences and authors, in both oral/aural and literary media, that allowed for allusive and lyrically economical communication of inherited material. Modern readers, less qualified as parties to such a contract, often miss non-explicit elements in an ancient rendition. Tithonus' various literary appearances provide a case in point as some scholars, noting that the detail of his transformation, previously missing from the record, first emerges long after Homer's time and Sappho's, conclude that the mutation motif is a secondary accretion to his story. That motif, though, even if unexpressed, is surely intrinsic to the overall narrative with its Dawn-Tithonus-cicada nexus. The myth itself could hardly exist were not the congress of Dawn and Tithonus a function of nature's timeless *son et lumière*, the matutinal coincidence of renewed cicada chorusing and the return of daylight. It is, in other words, the concurrence of audible and visible phenomena in the East that make Dawn (alias *Hemera*, Greek for "Day") and Tithonus the *tettix* natural partners, perennially, indeed eternally. The point is reinforced by Tithonus' name itself which shows the same sort of echoic or onomatopoeic repetition of apical stop consonants (Greek *th* being an aspirated stop as in English *anthill*) that *tettix* does. Compare, too, modern Greek *tzitzikas* and ancient Greek *stitthon* (used of another singing insect, the *akris*, or cricket), also the ancient verb *teretitzo*, often used of the cicada's voice. To call a character "Tithonos," then, would be phonetically analogous to calling him "Tettix." This matter of phonetic imitation of cicadas will recur in other contexts below. For the present, we will rest the case for Tithonus' name evoking the whole mythologem even as early as Homer to whom Tithonus now leads back.

A conspicuous stylistic device, first evident in archaic Greek poetry and frequently in later Greek and Latin poetry, is the extended "epic simile," also known as the "Homeric simile." Whenever these vignettes interrupt a narrative they simultaneously illuminate the specific topic of the context but also divert attention towards one or more tangentially related perspectives. It is within such windows in the narrative that most of the insect-related matter of the *Iliad* and the *Odyssey* appears, giving the insects an expanded, descriptively elaborate, poetic function. A salient example occurs early in Homer's *Iliad* as preparations are made beneath the walls of Troy for a decisive duel between the Trojan prince Paris and the Greek king Menelaus, respectively the current Trojan husband of Helen and the Greek husband from whom Paris seduced, or abducted, her. Arrangements for the duel interrupt an encounter of the Greek and Trojan armies which, we might assume, starts at dawn as Homeric battles often do. Spectators from both armies are present to observe the climactic fight (initiated but never concluded) while at the same time another assemblage of observers gathers on the city walls in company with the Trojan king Priam. It is at this point, with the instigators and principals of the Trojan War congregated, that the narrative leads into one of the more memorable of Homeric insect similes.

> Stationed on the Scaean gate was a group of elders of the community – Priam, Panthous, Thymoetes, Lampos, Klytios, Hiketaon (familiar with war) with their retinue; also Oukalegon and Antenor, a sapient pair. Being elderly, they had retired from warfare, but they were eloquent speakers, these Trojan leaders sitting on the tower just as cicadas sit and from a tree disperse their fluent song throughout the forest. On seeing Helen approaching the tower they calmly addressed one another with winged words. "It is no mark against the Trojans or the well-armored Greeks that they have endured long-protracted pains for such a woman. She is uncannily like the immortal goddesses in appearance and yet, precisely because of that, it is imperative that she sail back home and not remain as a cause of harm to us and our children hereafter."
>
> (*Iliad* 3.146–60)

This is among the earliest extant entomo-poetic passages in European literature, but it is likely to have derived from an ageless and familiar reservoir of entomological lore that enables the simile to extend beyond the explicit connection of cicada song and the discourse of the senior Trojans. It is therefore difficult to dismiss from the subtext the aged Trojan cicada-man Tithonus, the beloved of the Dawn as Homer elsewhere notes. Nearly all the elders on the tower—Priam, Thymoites, Lampas, Clytius, Hiketaon Antenor—here likened to cicadas of mellifluous voice, are identified in other literary contexts as Tithonus' brothers or, in the case of Antenor, brother-in-law. Not all auditors or readers would take away the same associations from the simile, but since chorusing cicadas in a treetop are a common fixture of pacific and idyllic poetic landscapes, the eloquent cicada-like elders conversing congenially with Helen and about her, present a stark contrast to the martial scene below. There are multiple, if perhaps inconsistent, hints of eroticism in the simile as well, for it is an observable reality and a commonplace of later poetry that the chorusing of male cicadas is sexually motivated. And here we have the *casus belli* herself, the preternaturally alluring Helen, in company with the cicada-like Trojan elders who can simultaneously marvel at her beauty and converse dispassionately about it in her presence, having withdrawn perhaps from the arena of Eros as well as that of Ares. In the meantime, on the other side of the simile, two armies are massed below in support of the erotically driven duelists. All things considered, the absence of the proverbially attractive Trojan prince and musical cicada-man Tithonus render him allusively conspicuous.

The simile of the cicadas and the Trojan elders occurs not quite a tenth of the way through the poem, but it has been preceded by two other similes that correlate the Greek army with insects. In the first the Greeks, energetically arousing themselves from nocturnal repose in preparation for the day's combat, are likened to bees that in angry agitation are stirred out of their resting place in a cleft rock. In the second, which comes several lines later within the same episode preliminary to battle, the Greek forces are being marshaled into ranks defined by ethno-geographic affinities. Here the poet uses an emphatic cluster of four similes to color in the numbers and diversity of the Greeks. They are likened in their myriads to the leaves and flowers in a riverside meadow, to flocking "tribes" (*ethnea*) of birds, differentiated as cranes, geese, and swans, to commingled herds of goats being separated by herders, and to numerous nations (*ethnea*) of flies congregating around milking pails in springtime. We might like to know if the word *ethnea* applied to flies would indicate a variety of insects, as it explicitly does in the case of the birds, or simply their vast numbers. What does become apparent here in this early stage of the *Iliad* is a pattern whereby insect imagery (here reinforced by reference to birds and springtime) serves to contrast the urgency and violence of the battlefield with the tranquil, quotidian, ambiance of pastoral and domestic life. There are other insect similes to support that impression. Among them is another application of the motif of the flies around the milk-pails where that side of the simile is echoed in a macabre variation. This time the soldiers correlated with the flies, while on the same battlefield where they were previously marshaled systematically, are seen teeming around the mangled and exposed corpse of Sarpedon, a heroic ally of the Trojans. Either instantiation of the milk-feeding flies is striking enough by itself, but for anyone who recalls the first upon encountering the second or who, on a subsequent reading of the passage, anticipates the second, the transformative effects of the imagery must be compounded. The insects, in short, without being directly involved in the events of the narrative, significantly texture its poetic environment. That said, though, it must be mentioned that there is one poignant scene in the *Odyssey* (which poem is more sparsely graced with insects than the *Iliad*) where the insects' role is not encased in a simile. One symptom of the general disorder that greeted Odysseus upon his long-delayed return to his home island was the situation of his old dog Argus whom he found, moments before his death, lying outside the palace amid heaps of manure, blind, neglected, and covered with vermin.

We cannot move on from Homeric epic without at least a passing notice of the Myrmidons, the military force led by the *Iliad*'s protagonist Achilles. The resemblance, quite possibly accidental in origin, of their name to *myrmex* the Greek word for "ant," predictably and typically supported an array of etymologizing explanations that involved ancestral ants. This etiology might well have been abetted by the observed behavior of warrior ants (Sears 2010). The *Iliad* itself in any case makes no reference to the explanation and indeed one of its epic similes alluded to above likens the Myrmidons to angry wasps swarming out to battle. Whether or not the poet is implicitly compounding insect imagery to suggest wasp-like ants, the linking of Myrmidons and ants is attested early in literary history, reportedly in a lost poem by Hesiod. Six or seven centuries later one version of the origin of the tribe was elaborated by Ovid, with all his sophisticated cleverness, in Book 7 of the *Metamorphoses*. Without any direct reference to Achilles or the Myrmidons at Troy, Ovid has Aeacus, Achilles' grandfather and king of Aegina, recount the origin of the Myrmidons: when the island was depopulated by plague, Aeacus prayed to Jupiter who responded with a show of lightning that, with suggestions of spontaneous generation, produced a multitude of ants, whereupon Aeacus requested a comparable number of warriors, and Jupiter complied by transforming the ants into the human Myrmidons. Ovid expatiates on the physical metamorphosis—the shedding of the ants' extra legs, the replacement of their slender shape, their assumption of an upright position—also on the admirable attributes of ants and Myrmidons without saying anything reminiscent of Homer's paradoxically wasp-like Myrmidons. Whether Ovid has reconfigured the myth himself, or simply selected from existing variant versions, he has ignored another account which has Zeus in the guise of an ant impregnate a Thessalian woman with a son named Myrmidon, the warriors' eponymous forebear.

Most of the once-copious epic poetry of Greece has been lost, leaving a gap of half a millennium between the archaic *Iliad* and *Odyssey* and the next fully surviving example of the genre, the *Argonautica* of Apollonius Rhodius. The *Argonautica* belongs to the markedly different poetic culture of the third century BCE that had evolved over the course of several centuries of post-Homeric literacy, but it maintains such formal devices of the genre as the Homeric simile of which Apollonius has several involving insects. One not only preserves the Homeric form but also recalls, even as it alters its tone, content from a specific Homeric context. When the Argonauts, on course for the Black Sea, reach the island of Lemnos they find it populated exclusively by sex-deprived women who later, the poet says, flew out to the heroes at the time of their departure in the manner of bees swarming out from a rock to dart from one sweet flower to another, all in a sympathetically joyous landscape. The women express their wishes for the heroes' happy return to Lemnos, but despite the ostensibly amicable parting, so different from the emergence of the bellicose Homeric insects, Apollonius' simile inherits ironically ominous overtones in the perception of any reader acquainted with the entire story of the Lemnian women. They, after all, had previously killed their Lemnian men-folk in retaliation for temporarily abandoning them and then returning to Lemnos with concubines captured in Thrace. What sort of welcome, under what circumstances, would the returning Argonauts really receive?

If any major genre of antiquity has only a modest cohort of insects it would be tragedy, both the fifth-century Athenian plays and those of the Roman Seneca over four centuries later. There is, though, one striking instance involving spider imagery. Early in the *Agamemnon* of Aeschylus, as Clytemnestra anticipates her husband's imminent homecoming with the unspoken intention of murdering him, she reports on a relay of fire signals announcing the end to the war at Troy. The sequence of signals concludes, ominously, at an elevation known as "Spider Crag," close to the palace at Argos where the king will shortly be murdered. The word translated here as "Crag" has a semantic history with overtones of death, darkness, and adversity in general, while the spider allusion anticipates the mode of Agamemnon's death in which Clytemnestra lures him into the palace and then entangles him in a fabric which, more than a thousand lines later, is twice explicitly called a spider's web. The arachnid imagery, then, with Clytemnestra as a spider-woman is sustained over virtually the whole length of the play. There can be little doubt that it carries sinister arachnophobic overtones. The same impression is left by a terse comment from the frantic chorus of refugee Danaids in *Suppliants*, another Aeschylean tragedy. While under threat from the pursuing Egyptians, the daughters of Danaus refer to the

Egyptians' herald as a spider whose terrifying approach is nightmarish (literally "a black dream").

Elsewhere tragedy does allow a prominent, also unpleasant, function to a true insect, the gadfly, as a player in the story of Io. She, one of Zeus's many human paramours, underwent a transformation into a comely heifer. The point of interest here is that Io was then beset by the jealous Hera with a tormenting gadfly that drove her from Greece, through the Balkans, across the Bosporus, and all the way south to Egypt. Io's references to her ordeal are conspicuous in the *Prometheus Bound*, a play attributed, possibly erroneously, to Aeschylus. The gadfly also joins the spider in the authentically Aeschylean *Suppliants* which includes a cursory reference to it.

Comedy presents a fuller and more varied entomological menagerie than tragedy (Conti Bizarro 2009). This is mainly apparent in the surviving plays of Aristophanes staged in Athens in the later fifth and early fourth centuries BCE. A formal feature of the genre in this period was a chorus of singers and dancers that complemented the actors. Such choruses were frequently costumed as animals, sometimes as insects, and reflected in their anthropomorphized words and actions the ethology of those creatures. We know of now-lost plays with titles such as *Ants*, *Ant-men*, *Bee*, and *Bees*, and one of Aristophanes' eleven extant plays has a chorus of *Wasps* who gave the play its title. The characteristic irascibility and aggressiveness of the wasps with their punitive stingers accommodate their representative role as jurors in the Athenian courts. In that respect they exemplify Athenian comedy's characteristic use of ribald absurdity in criticism and satire directed at social, political, ethical, and ideological targets. In the case of *Wasps* the entomological intrigue seems to extend into literary or meta-theatrical criticism as Aristophanes intertextually plays off his cantankerous and adversarial wasps against the socially beneficent bees associated with the personified *Laws* of his rival playwright Cratinus (Sidwell 2009: 186–8).

Each of the other comedies of Aristophanes in some memorable way exploits insects for comic effect. A thorough description of the whole inventory with its bed bugs, flies, moths, lice, cock-chafers, etc. would fill many pages (Harris 1989), so a few salient examples must suffice. In *Peace*, presented in 421 BCE when the Greek states were about ten years into the Peloponnesian War, a pacifist named Trygaeus appears on stage astride a horse-sized dung beetle (recalling the mythical Bellerophon on the winged horse Pegasus) that flew him to the celestial home of the gods in successful quest of a divinely arranged end to the war. Typically for comedy of the era, the beetle-back ride is colored with appropriate scatological atmospherics of sound and odor. A beetle of a different type is another object of preposterous ingenuity in the *Acharnians* where a scheme is mooted for sabotaging the Athenian fleet by having a water-beetle float into the dockyards attached to an incendiary candle. The *Clouds* presents much spoofing of the intellectual speculations and scientific investigations of the day. It seems, indeed, that entomological studies, such as those reflected in the following century by the sober prose of Aristotle's zoological writings, were among the targets here of Aristophanes' satiric mockery. There is a scene, at any rate, in which curious investigators encase the feet of a flea in wax booties that will mark the point where the leaping insect alights, thus enabling the precise measurement of the span of the leap. *Clouds* also presents entomological theorizing in the form of a pseudo-sophisticated discussion of the anatomical niceties of the mosquito's sound production. The conclusion is that air inhaled through the insect's mouth passes through its innards to be expelled with trumpeting sonority from its rump. Another Aristophanic example comes from a play called *Wealth* in which a formerly impoverished farmer named Chremylus dismisses Penia, the goddess of Poverty, from his home where she is to be replaced by Plutus, god of Wealth. Chremylus charges Penia with harboring such annoying and noxious cohabitants as lice, mosquitoes, fleas, and bed bugs. When Poverty voices an entomological counterclaim by noting that she makes the men with whom she associates waspishly thin and effectively antagonistic to their enemies, Chremylus retorts that their wasp-like physique is caused by starvation.

The extent of the insect population of Aristophanes' surviving works indicates that the loss of the great majority of the plays of Old (i.e., fifth century) Comedy is a particularly regrettable one for literary entomophiles. This is confirmed by sporadic evidence in the form of fragments from lost plays preserved, mostly in citations by later writers. One example is a lyrical passage of about twenty-five words from a play of unknown title by an author named Strattis. The passage addresses an insect called a "leek-cutter," apparently a caterpillar, who dances among garden vegetables with his fifty feet. In a later era of theatrical history, when Greek comedies provided models for Latin translations and adaptations, the Roman playwright Plautus hints in a comedy entitled *Curculio* (Latin for "weevil") that he might be adapting a lost Greek play with an insect cast. Appropriately, the title character Curculio is a parasite, that is a freeloader of the type that is a stock character of the Roman comic genre and its Greek antecedents. While the aptness of Curculio's name has always been recognized for its implication of voracity, it has largely gone unnoticed that Plautus' dramatis personae also include a predatory banker named Lyco (the Greek term for "wolf spider") whose principal victim is a pimp explicitly likened to such spider fodder as flies and mosquitoes as well as other vermin such as fleas, lice, and bed bugs. The entomological coloring of the play, moreover, is all applied within a plot that features the amatory quest of a character named Phaedromus who, by the light of a candle provided by little "bee-lets," goes out in search of his "sweet little honey." Said little honey has fallen into the hands of the pimp who subsequently defaults on delivering her to a formidable soldier who then furiously threatens to pulverize the hapless pimp so that ants will carry him away, crumb by crumb.

Without flying very far from the realm of comical insects we can now entertain two examples, both from the first century CE, of literary parody with an insect protagonist. One of these is from the pen of a prolific satirist, a Hellenized native of Syria named Lucian, whose oeuvre includes humorous variations on established literary forms. In this case, the model imitated is a type of rhetorical exercise that was part of the professional training of teachers, politicians, and lawyers or any others whose ambitions called for well-developed oratorical skills. These exercises often entailed composing encomia or laudatory speeches for historical or mythical persons, animals, or even objects. This is the genre in which Lucian amusingly extols the anatomical, vocal, ethical, and spiritual attributes of the fly, often pejoratively citing other insects—grasshopper, gnat, bee, locust—in comparison. The comparisons indicate an accurate familiarity with physical and ethological characteristics of both the fly and the other insects, and much the same must be said of the author's acquaintance with literary entomology. He approves, for instance, Homer's representations of the fly's courage, but criticizes Plato for not recognizing that, like humans, the fly has an immortal soul. A miscellany of other commendable points would include the extended duration of the fly's copulation, its ability to copulate in flight, its wings as beautifully vari-colored as a peacock's feathers, its cleverness in avoiding spider traps, its role as a dinner guest at the tables of kings, and its dextrous use of its two front legs in the manner of human hands.

The second parodic piece is a Latin spoof of a type of short narrative poem, a miniature epic sometimes labeled *epyllion*. The anonymous *Culex* (Latin for "gnat" or some other small biting insect), as it is entitled, has come down in a collection of shorter poems once attributed to the younger Vergil. In *Culex,* though, genuine works by Vergil are in fact among those parodied. These include the epic *Aeneid* and his shorter pastoral poems, the *Eclogues*. Like many *epyllia*, then, the *Culex* incorporates conventional devices of both the epic and pastoral genres. Relevant to the present context it might be noted that it includes a passing reference to a cicada, a de rigueur feature of the pastoral landscape in both Greek and Latin poetry. The tragic hero of the piece, though, is the gnat who bites a sleeping shepherd, thereby awakening him in time to avoid a lethal attack from a serpent. Before bludgeoning the serpent to death, however, the shepherd has reflexively swatted the life-saving gnat and sent it prematurely to the afterlife. From there the gnat, miffed at the unthinking ingratitude of the shepherd, later appears to him in a dream in which he voices his complaint but also gives a full-blown, epic-style, description of the sights and the mythical personages encountered in the underworld. The chastened and now grateful shepherd then builds and dedicates a shrine replete with a commemorative inscription to the insect hero.

The idea of an epigraphic tribute to an insect is hardly in itself an invention on the part of the *Culex* poet, for we know of several poetic epitaphs for insects such as cicadas, crickets, and grasshoppers. Perhaps some such epitaphs were once actually carved on grave markers in honor of pet insects, but it is likely that many of them were purely literary diversions composed in the traditionally lapidary format of epigrams that really were incised in stone. The literary epigram endured for centuries as a miniature poetic form often consisting of a single couplet and seldom of more than seven or eight couplets. Its prescribed brevity presented special challenges to the poet's resourceful ingenuity. On the Latin side, the pre-eminent representative of the genre is the large oeuvre of Martial, a miscellany of over 1,500 separate poems marked by sardonic wit and irony. Martial, in keeping with long epigrammatic tradition, includes insects in his repertoire. One four-line poem, for instance, muses on an ant, a creature disdained during her lifetime, who in death became a highly valued object encased in amber after being caught in gooey resin. In quite a different mode, and in a departure from epigrammatic convention, Martial opens twenty-five lines of iambics (the traditional meter of satire and invective) with an apparently pitiless assault on an aged, virtually bald and toothless (having three hairs and four teeth) prostitute with the surely fictional name of Vetustilla (i.e. "Old Girl"). Ironically, though, he closes the poem with a hint of pathos, noting that her next liaison will be with Death. His poetic volte-face comes only after a catalog of derogatory comparisons to various animals in which he gives her the chest of a cicada, the shanks and complexion of an ant, breasts like spider webs, and a singing voice inferior to a gnat's. In another of his longer poems Martial similarly accumulates insect imagery in a complaint about the minuscule size of a rural property gifted to him; it is too small to shade a cicada, to feed an ant for a day, or a single caterpillar in its garden, and a gnat would starve to death there.

Antecedent to Martial and Latin epigram in general there was a rich Greek tradition that is documented in works from as early as the sixth and seventh centuries BCE. That same tradition continues unbroken down through the Byzantine period and beyond. It has left a massive monument in the form of what is now known as the *Greek Anthology*, a collection of thousands of poems by scores of poets composed over many centuries and compiled incrementally through several revisions over the course of a millennium. Its significance here is as a rich repository of entomopoetic artistry, but consideration of representative poems must be introduced by notice of a few general features of the anthology's history and the character of its contents. If sepulchral, commemorative, and dedicatory or votive themes were dominant in the earlier phases of the tradition, they were soon joined by the amatory, the pastoral, the philosophical and the ludic. Frequently, several such themes were blended into single epigrams, many of which involve insects. Poet-editors had been compiling collections of epigrams of their own works and/or those of others for several generations before the mid-first century BCE. At that time a Hellenized Syrian named Meleager from Gadara (at the time a major center of Greek culture located in what is now Jordan) compiled an anthology (literally a selection of "flowers") containing epigrams composed over the course of several centuries by some forty other poets each of whom he identifies with a particular flower. This collection of poetic blossoms was woven together with epigrams by Meleager himself into a *Wreath* (*stephanos* in Greek; commonly also translated as *Garland*). This compilation remained a major contributor to later anthologies of epigram, although Meleager's complex verbal and thematic interlacings have been interrupted and obscured (but not completely obliterated) in the long process of accretion and editing that produced the *Greek Anthology* as we have it from Byzantine times. Since Meleager is a key figure in the evolutionary history of the ancient epigram, an editor and preserver of earlier insect poetry, and a model emulated by later epigrammatists, it will be appropriate to adopt him as the epitomizing focus of attention to the insects of the *Anthology*.

To demonstrate the combination of the ludic and amatory, a pair of eight-line epigrams addressed to mosquitoes will serve. Their juxtaposition in the *Anthology* in its present form, if Meleager is responsible for it, carries a certain irony; after mosquitoes appear as pests in an erotic context in the first poem, one of them is invoked as an aide in the amatory intrigue of the second poem. In the first, the lover of one Zenophila invites the mosquito to allow his beloved to sleep by feeding on his flesh rather than hers, but he soon concedes that those ruthless beasts would find her skin preferable to his. The poem closes as he threatens the mosquitoes with physical violence. While the same Zenophila and a mosquito are principals in the second epigram, there is a marked change of sentiment towards the insect. For one thing, the insects of the first epigram are introduced as "shrilly shouting," while the addressee of the second is flattered as a "lover of music" or "beloved of the Muses" and the words of the opening couplet are marked by an imitative susurration of sibilant sounds as if the lover is buzzing in the vocal medium of his mosquito confidant: *ptaiē***S** *konop***S** *tachu***S** *angelo***S** *oowa***S***i dakroi***S** *Zēnophila***S** *p***S***au***S***a***S** *pro***S** *p***S***ithurip***S**. The communication proves effective as the hopeful lover's whispered Greek instructions send the mosquito off to whisper in Zenophila's ear the message that he is ready and waiting for her. The mosquito is enjoined to speak softly to avoid arousing the rival lover who is sleeping beside Zenophila. This epigram concludes with a mock-heroic touch, comparable to the slightly later heroization of the gnat in the Latin *Culex*, when the mosquito is promised a club and a lion-skin helmet, traditional emblem of Hercules, as a reward for a successful and heroic mission.

A more complicated Meleagrian example of insect-inspired virtuosity centers on another pair of epigrams, juxtaposed in the existing *Anthology* as they almost certainly were in Meleager's *Wreath* (Egan 1988). Inevitably, much of the verbal artistry does not come across in translation but the following rendition will provide a basis for comment.

O cricket, reliever of passions through the solace of sleep;
O cricket, shrill-winged rustic Muse,
Nature's mimic of the lyre, chirp for me a plaintive song,
Strumming loquacious wings and tapping your endearing feet,
Allaying oppressive, sleep-depriving, torment while,
O cricket, you dismiss Eros with continuous song.
I shall give you as presents at the rising of dawn
A fresh leek and mouth-sliced dewy droplets.

O resonant cicada, drunk on dewy droplets,
You sing your rustic song that sounds in the solitude.
Perched high upon the leaves, on saw-like limbs,
With sun-tanned body you shrill forth a lyre's tune.
But, my friend, voice something new for the woodland nymphs,
A playful tune, as you chirp a responsorial song to Pan,
While I, escaping Eros, pursue a mid-day sleep,
Reclining there under the plane tree.

Conventional readings presume in each case that a human is addressing the insect. This at least yields a pair of variations on the theme of insect song providing a human being with consolation and respite from the labors of love. All of that would be consistent with the age-old epigrammatic challenge to rework an existing poem, simultaneously imitating it and competing with it, although such competition usually involves more than one poet. There is, though, another way of reading this matched pair that recognizes them as part of a poetic and editorial tour de force in the application of epigrammatic conventions.

In the first poem the cricket is addressed at close of day and asked to begin his song, just as crickets do at that time which, by nature's synchronization, coincides with the time when the cicada goes silent and retires from his daylight musical and erotic labors. Those pursuits the cicada will resume at dawn when he also compensates the cricket for his soothing nocturnes with a refreshing breakfast. It is, in short, not the voice of the poet nor of any human persona that addresses the cricket here, but that of the weary cicada. In the second poem the roles are reversed with the cicada providing a daytime lullaby while the exhausted cricket dozes. The two pieces are linked not only by the reciprocity of their conceits and vocabulary but also, at the boundary between them, by a striking piece of aural imagery. It would be futile to attempt a translation that replicates the harmonization of semantics and sound achieved in the final line (A) of the first poem and the opening line (B) of the second, but the following transliteration is an approximate representation of the Greek sounds.

(A) *kai dro**S**era**S** **S**toma**S**i **S**chi**Z**omena**S** p**S**akada**S***

(B) *achēei**S** tettik**S** dro**S**erai**S** **S**tagone**SS**i methu**S**thei**S***

The crescendo of nine sibilant fricatives (marked by bold capitals) that closes the first poem is sustained with another nine at the opening of the responsorial second poem, other features of note include the occurrence in each line of a phonetically similar grammatical variant—*droseras* and *droserais*—of the Greek word for "dewy," each of which modifies a semantically and syntactically matching word for "droplets"—*psakadas* and *stagonessi*. All of the echoic effects here are achieved, not by using imitative nonsense words, but by composing intelligible and metrically correct Greek which is also entomologically multi-referential. The effect of the insects' *Lautmalerei* at the boundary between the two poems is reminiscent of the sound effects in Meleager's second mosquito poem but the general technique was not unique to Meleager. We shall shortly find it employed in the poetic prose of quite a different genre almost four centuries earlier.

The brief repeated references to dewdrops in the adjoining lines of the two epigrams allusively compress additional entomological lore and realia. It was a popular belief that cicadas subsisted on dew and an observable fact that they draw fluids from trees by puncturing the bark, thus allowing some fluid to escape and moisten the trees and anything beneath them. The fluid that they imbibe in considerable quantities, moreover, they also excrete in comparable profusion processed as what is now called "honey-dew," a sugary liquid known to be consumed by other insects such as ants. Therein lies a likely explanation for the intoxicating droplets consumed by Meleager's cicada and also those, separated by his mouth, with which he seasons the leek for presentation to the cricket.

Lengthy though it is already, the explication of Meleager's exchange between the cicada and the cricket is not yet complete, for not only has he woven his two epigrams together verbally and thematically, he has also embedded the composite in a larger context. In the *Anthology* as now constituted Meleager's own sixteen verses of the cicada-cricket pairing stand enwreathed, or entangled, in a cluster of poetic epitaphs for cicadas and crickets previously composed by other poets. Among those epitaphs, nine in number, there is one belonging to a single tomb shared by a cricket and a cicada. In three others the insect itself proclaims its own epitaph. The nine epigrams together amount to a total of forty verses of which twenty precede and twenty follow the Meleagrian pair. The sequence and its symmetry are broken on each side at an interval of twelve verses following Meleager's contribution by a bird epitaph which, being anomalous in subject and wording, might have been inserted by a later anthologist. In any case, the Meleagrian set of nine plus two cicada-cricket poems in a total of fifty-six verses is unified by more than their shared insect subjects, for Meleager has bound the several assembled epigrams together by elaborately intertwining in his own poems numerous key vocabulary items coopted from his predecessors. The ensemble when read as a unit must be acknowledged as a miniature masterpiece of entomopoetic composition (Egan 1988, 1998).

Another virtuoso composition featuring insects in life and death figures conspicuously in one of Plato's philosophical dramas, the *Phaedrus*, probably the most artistically complex and eloquent work of the Platonic corpus. In another dialogue, the *Phaedo*, Plato has Socrates propose that human observers of justice and moderation might be reincarnated into a society of ants, bees, or wasps, but in the *Phaedrus* a kindred notion of humans being transformed into cicadas (recall Tithonus) is fully elaborated by Socrates. Plato carefully sets the stage by having the interlocutors, Phaedrus and Socrates, meet outside the city where they agree to discourse on eros in an idyllic rustic setting while seated at noontime on a grassy riverbank beneath shade trees in which cicadas sing (Egan 2004). The circumstances and themes amount to a catalog of stock motifs of pastoral poetry as found later in works by such poets as Theocritus and Vergil. The cicadas turn out to be not merely an incidental fixture of the Platonic setting but a factor integral to the philosophical themes of the dialogue. In a variation on a Tithonus motif, Socrates explains that they were once humans so devoted to the Muses and music-making that they forgot to eat and were transformed into cicadas. Henceforth they serve as intermediaries between humans and the Muses including, pointedly, those with special responsibility for erotic and philosophical themes. A corollary to that last detail has Socrates proclaim that these prophets of the Muses have inspired his two speeches on eros. As if to underscore that, Socrates is actually represented sounding like a cicada while speaking correct, sense-bearing Greek. Consider, for instance, the following sequence of transliterated Greek in which virtually every consonant is an apical (in bold type) stop, plosive, or fricative, as are those of *tettix* and *Tithonos.*

> **theos-ōn-theios-estin-tois-de-dē-toioutois … teleous-aei-teletas-teloumenos-teleos-ontōs**

There are, then, varied and redundant indications that the cicadas are active and inspirational participants in the dialogue which includes Socrates' elaborately figurative description of the development of the philosopher's soul that culminates with its maturation as a winged being after having left both its constricting body and the subterranean darkness. That description reflects in many of its details the final stage of cicada metamorphosis when the wingless nymph emerges from beneath the ground, sheds its integument, and deploys its wings in a sequence that, as Aristotle indicates in some detail, was well observed by the ancients.

The matter of emergent cicada-souls gives rise to a couple of tangential points that lead temporarily beyond the *Phaedrus*. The notion that cicadas were earth-born led the Athenians to regard them as symbols of their own autochthony. The same notion appears to be operative in *Symposium*, another Platonic dialogue on eros, when the playwright Aristophanes as one of the interlocutors in Plato's work delivers a comical account of the evolution of human reproduction that primordially involved birth from the earth.

The subject of the insect-soul also prompts a brief digression here on Lepidoptera which, as ancient lexicographers attest, and visual representations corroborate, were in their adult form as moths or butterflies regarded as souls with which they shared the designation *psyche* (Davies and Kathirithamby 1986: 99–108; Beavis 1988: 121–31; Icard-Gianolio 1994: 583). They are mentioned parenthetically here for the curious fact that they have virtually no explicit presence in ancient literature (as opposed to the visual arts) despite their attractive and conspicuous ubiquity in human experience. The reason for their neglect or avoidance by writers is unclear although one suggestion attributes it to a taboo connected with the popular superstition that they are the souls of deceased humans. Whatever the explanation, we can surmise that the concept was at least potentially implicit in instances where the linking word *psyche* occurs. It is natural, for instance, to suppose that at least some readers of the *Phaedrus* will have found the Platonic imagery of the cicada/psyche vocative of the butterfly/psyche.

Plato assigned another major role to insects in his *Republic*, one of the most famous and influential documents in our intellectual history. Here, too, as with the cicadas of the *Phaedrus*, Plato works within a long poetic tradition from which he creatively deviates in the service of philosophy. This time the paradigmatic insect is the bee, for centuries a symbol and metonym of poetic skill and inspiration involving the analogy between the sweetness of honey and the sweetness of poetry. A catalog of poets and critics who, before and after Plato, encomiastically embraced such favorable treatment of the bees would read like a *Who's Who* of ancient poetic history. Plato himself reflects the tradition in a passage of his *Ion* where Socrates observes that poets claim an analogy between their own gathering of sweet inspiration from the Muses' haunts to the honey-gathering of the bees. Socrates agrees with the poets' claim, but hardly approvingly because he regards poetic inspiration as irrational. In the *Republic*, though, Plato is more direct in his criticism of the entomoidal poets as he emphatically subverts the complimentary tradition by focusing on the non-contributing drones of apian society and comparing them to the poets whom, notoriously, he deprecates for their deleterious effects on human society (Liebert 2010).

Approximately six centuries later an engaging sequel to this part of the *Republic* and indeed to the cicada sections of the *Phaedrus* appeared in *On the Nature of Animals*, written in Greek by a learned native of Palestrina in Italy named Aelian (Smith 2014). The work abounds in insect fact and lore, much of it derived from earlier writers such as Aristotle, but Aelian takes considerable initiative when he writes about bees vis-à-vis Plato. He tries to rehabilitate the society of the bees to their pre-Platonic position as worthy models for emulation by human societies. He reacts, apparently deferentially, to Plato's comparison to the corrupting social effects of drones and poets, by admitting the need to neutralize the negative drone elements in society. Less deferentially and perhaps a bit disingenuously, though, he counters the *Republic*'s aversion to poets by citing the myth of the cicadas from Plato's own *Phaedrus* for its evident approval of the work of the Muses.

Socrates is a versatile vehicle for Plato's varied applications of insect figurations. If other qualities and contexts fitted Socrates to cicada imagery, his stimulating and at times aggravating habits of incisive inquiry suited him for metaphorical identification with the gadfly in *Apology* where Plato seems to supplement the gadfly's attributes with others borrowed from flies and wasps (Harris 2018).

By now a cluster of mostly benign notions about insects, notably bees and cicadas, as incorporeal entities—divinities, divinized humans, souls free from bodily constraints—has been observed coursing through ancient literature from its beginnings. Three additional selections, salient for different reasons but all reflecting philosophical or religious concepts of spirituality, will augment the list. A sequence in Vergil's *Aeneid* exploits current ideas about the transmigration of souls between life and death when the hero Aeneas visits the underworld to preview the future of the nation he is destined to found (Horsfall 2010). Aeneas sees the souls of past acquaintances and of individual heroes and villains awaiting reincarnation as eminent figures in the future history of Rome. He also has a vision of innumerable bees who represent the souls of humans in the blissful environment of Elysium, the area of the underworld reserved for those who previously led a righteous life. The second selection is an anonymous ode from an unknown date in the early centuries CE and included in the *Anacreontea*, a poetic miscellany falsely attributed to Anacreon, a poet belonging to a much earlier era. The piece has gained some modern attention through eminent translators such as Goethe, Cowper, and Thomas Moore. Formally it is an example, likely a parody, of the *makarismos*, that is a declaration of blessedness or felicity common to Greek and Latin religious literature, pagan and Christian, from the Archaic period on. For present purposes a prose translation will serve as efficiently as any summary to demonstrate that the poem is also a synopsis of common literary features of cicada ethology.

> We regard you, o cicada, as blessed as a king when, having drunk a little dew, you sing from the treetops. For yours are all the things that you survey in the countryside and all the things the woods produce. You are beloved of farmers and do no harm to any one of them. You are esteemed by humans and the Muses love you as their mouthpiece in the summer, while Apollo himself also loves you and has endowed you with a clarion voice. Old age does not oppress you, you artful one, earth-born, song-loving, immune-to-suffering, flesh-and-bloodless. You are virtually the equal of the gods.

In some ways reminiscent of the Anacreontic ode is another religiously inspired adaptation of cicada motifs in the form of a Greek Easter-tide sermon composed by one Asterios, probably in Syria in the late fourth or early fifth century. Asterios draws several analogies between cicadas and Christ or Christians: Christ on the cross is like the cicadas in the tree-top, the dew-imbibing cicadas are like the Christians partaking of divine wisdom; the emergence and wing deployment of the cicadas resembles the resurrection of Christ, but also baptismal rebirth; the white-clad candidates for baptism resemble the initial whiteness of the adult cicadas, etc. We leave Asterios to serve as the token representative of numerous early Christian authors, both Greek and Latin who, more briefly than Asterios, exploit the figurative potential of insects (Davies and Kathirithamby 1986: passim; Beavis 1988: passim).

More must be said about Vergil, the *non pareil* of Latin writers for the quantity and sophisticated artistry of the entomological poetry dispersed throughout his three surviving works: *Eclogues*, a collection of ten poems in the pastoral mode; *Georgics*, a set of four poems, didactic in mode and ostensibly devoted to agricultural matters but, through allegory and allusion, intricately related to ethical, political and social concerns; and *Aeneid*, an epic that honors formal conventions of the genre as first represented by Homer while dealing with philosophical, religious, social, military, and political matters related to the origins, history, and current situation of the Roman state. Being pastoral, the *Eclogues* contain generically appropriate references to bees and cicadas. The fourth book of the *Georgics*, all 566 lines of it, is in a sense entirely entomological, being devoted to apiculture, however much that theme might be attenuated by historical and political matters or by mythological digressions such as a lengthy version of the Orpheus myth. That narrative is linked to lines on the *bugonia*, the putative process by which bees were spontaneously generated from bovine carcasses. To put it in the briefest of terms, the poem uses the society of bees as a paradigm of human society, a point that is made clear in a sequence of lines on the division of labor in apian society. All of Book 4 has also been anticipated with several lines on bees within an overture section in Book 1.

Early in the *Aeneid* Vergil intertextually retrieves lines of his own on the collaborating bees from *Georgics* 4 when the hero Aeneas, newly arrived in Africa, views from a distance the Phoenician colonists working in the fashion of bees to construct the new city of Carthage (Giusti 2014). Because there would be readers who recognized his self-quotation of the earlier linking of the bees to contemporary Roman society, the same complex simile of the bees would also represent Aeneas and the Trojans as the destined founders of the Roman nation and, moreover, it would recognize that at the very time of the *Aeneid's* writing, the bees (now as Romans) were building Carthage anew, this time as a Roman city. The bees of this part of the *Aeneid*, in other words, are metaphorically occupied in the foundation of the three historical cities: Phoenician Carthage, Rome, and Roman Carthage. Besides all that, they are linked with another complex simile based on other social insects, the ants. In effect, the long Carthaginian episode of the *Aeneid* is bracketed by two multi-referential insect similes since, in Book 6, as Dido looks down from her funeral pyre at the departing Trojans methodically lading their ships with provisions, they are seen as ants at work. From Dido's perspective, the ant-like Trojans are plundering Carthage (as the Romans did, centuries later); from the Roman readers' perspective, their ant-like Trojan ancestors prefigure their contemporaries who, in Vergil's own time, industriously exploit the natural and agricultural resources of what had become Roman Africa.

In keeping with the comparison suggested earlier likening classical literature to a complex organism, it seems appropriate to take note of a piece of insect-inspired advice from the pen of Lucius Annaeus Seneca, the Stoic essayist, satirist, and playwright of the first century CE. A principal theme of one of his epistolary essays addressed to a correspondent named Lucilius (84) is the artistic and intellectual value of complementing writing with wide and eclectic reading (Berrens 2015). To that end, he presents the foraging of bees and their production of honey as a model for the successful reader-writer's metabolizing of varied literary gleanings into something original and distinct from its individual ingredients. He refers to the long antecedent poetic and meta-poetic tradition connecting apian and poetic creativity, even citing Vergil's simile of the bees at Carthage. He also allows himself a few words on competing theories of honey production—do bees collect honey ready-made or do they re-constitute the collected ingredients? Ironically, that question might be asked of Seneca's own parable because it has a long string of antecedents in Greek and Latin literature. Notable among those is the poetic profession made a century earlier by the eminent apostle of the Epicurean philosophical school who compared himself to honey-gathering bees, with the qualification that he restricts his browsing to the meadows of Epicurus.

Those meadows themselves, though, had previously accumulated eclectically the wisdom of Epicurus' philosophical predecessors. If Seneca's allegory of the eclectic bees has a long history behind it, it has also had a fortuitous role in the still-lengthening reception history of the insects of ancient literature, for it was recognized in the Renaissance as a figurative recipe for Humanistic learning (Falco 2017: 148–51). Much more recently still, it has been recognized in blogs as a pioneering manifesto for scientific combinatorics.

We are now nearing the conclusion of this representative survey of the insects that populated Latin and Greek literature over many centuries—buzzing and biting, foraging and pillaging, flitting and twittering, trumpeting and whispering, while exemplifying a wide and varied spectrum of human wisdom and human vanities. The repeated cultural resurfacings of Seneca's exemplary bees have also just readvised us that the early history of cultural entomology is an ongoing process. This, then, is an appropriate juncture for a general observation about the enterprise. Ideally, all data and interpretations such as those presented with some confidence above are distilled from centuries-old Greek and Latin texts through the labors of modern readers with appropriate competence in the norms and niceties of those languages, in the various artistic and intellectual times and circumstances that generated those texts, in the generic conventions among which the texts are deployed, and with the imaginative and artistic habits and devices of their authors. Those conditions must be matched with the biological and behavioral functions of insects as observable, and particularly as they are known to have been observed in ancient times. This all amounts to a complicated recipe in our times, and one that is frequently ignored by contemporary translators and commentators, Latinists and Hellenists, who in our urbanized, mechanized, artificial, entomophobic, and insecticidal environments might be denied the quotidian opportunity that the ancients apparently had to familiarize themselves with the activities of insects and reflect them in their literary culture. A salient example of resulting confusion, albeit not a particularly modern one, is the common transmutation of the cicada into a grasshopper via translations of the Aesopic fable of the lazy cicada and the industrious ant. That practice might be mitigated by translators' intentions to present young readers with the visibly more familiar grasshopper in place of the commonly camouflaged cicada. There is, though, no such redemption for contemporary translators and commentators on classical literary works to identify cicadas as grasshoppers, or to confuse their integuments misleadingly with cocoons or chrysalises. Ironically, modern technology, together with the efforts of professional and amateur entomologists, has provided a corrective in recent decades. It would be a rare, non-specialist, reader of this chapter who, without laboring in the field or laboratory but remaining at a desk with a computer, could not view or hear a proliferating abundance of insect activity courtesy of the internet. Some such consultation might indeed afford some helpful enhancement to the following prosaic discussion of the final passage of ancient literature to be entertained in this chapter.

There follows below a translation of an excerpt from that section of Lucretius' poem on the origin and functions of the universe in which he presents his adaptation, consistent with the philosophical teachings of Epicurus, of traditional speculation on zoogony, the primordial spontaneous generation of animal species from the earth's interior. The passage recapitulates theorizing that can be traced back over several centuries to the now sparsely and randomly preserved writings, some of them poetic, by Greek thinkers of archaic times such as Empedocles. As in the entirety of his poem, a challenge taken by Lucretius here was to fit concepts that he received from Greek predecessors to the novelties of the Latin language and its still evolving poetic techniques. Anyone familiar with the passage, whether in Latin, or in any modern translation or commentary, will find this version idiosyncratic in several respects, but it is consistent with the prescriptive recipe presented above (I intend to publish a more detailed explication with direct attention to lexical and stylistic features of the Latin text and some Greek antecedents), and the formatting anomalies have a purpose.

> Earth has deservedly gained the title of Mother because all creatures come out of the earth. Even **now** many living beings come into existence from the earth after being formed by rain and the warmth that emanates from the sun, so it is hardly surprising if, **at that time**, when both the earth and the atmosphere were young, more numerous and larger creatures arose and flourished. To start with, the class of winged creatures, including the birds in their variety, began evacuating and leaving their eggs in springtime just as **nowadays** the smooth cicadas spontaneously leave their wombs (Latin *folliculos*) in quest of nourishment and life so **at that time** Earth first gave forth generations of creatures destined to die out, since an abundance of both heat and moisture prevailed in the fields. For that reason, wherever a particularly suitable location presented itself, wombs (Latin *uteri*), attached to Earth by roots, would grow and, when the time was ripe and the infants' life-stage had forced them (i.e. the wombs) open, their natural impulse, avoiding the moisture and questing for air, would thereupon direct them (i.e. the wombs) towards perforations in the earth and prompt them to pour out from open veins a juice, comparable to milk, just as **now** every female, after giving birth is filled with sweet milk because the entire thrust of her nourishing function is directed towards her breasts.

The excerpt has been divided into three sections. The shortest of the three containing the only explicit reference to insects is here printed, unconventionally for *modern* practice, without any punctuation before or after. By itself, that section is not a complete syntactic unit in either English or Latin. The normal practice of modern translators or editors of the Latin text is to place a full stop after that short segment, thus connecting it syntactically with the clause at the end of the first segment and isolating it from the clause at the beginning of the third segment. That completes a relatively simple simile that links the ecdysis of cicadas with the hatching of birds, and that is where modern readers let it end. If, however, a full stop is placed at the end of the first segment, the second segment must connect with the third one, thus joining in a more complicated simile resembling the Homeric type. If we must have a full stop in one or the other location, since either one would yield an intelligible and relevant simile, we have a parsing dilemma or ambiguity. It is, though, quite likely that the original Latin text of the first century BCE did not mark a stop in either position, thus allowing for a case of double syntax hinging on the trope of the emergent cicadas. It seems, then, that the poet, experimentally creative as he is in fitting the Latin language and his message to each other, has in an almost literal sense made the cicada a pivotal figure in his speculation on the primordial origins of animal species.

For the purpose of pursuing that latter point, some words in the translation have been bolded to emphasize how the argument uses the proverbially earth-born cicada of the present as a sort of living fossil, surviving proof as it were, of the prehistoric spontaneous generation of all species. The references to the "wombs" (expressed in the excerpt with two different Latin words) is a known inheritance from Archaic Greek speculations on zoogony, but the conceit here related to cicadine ecdysis must be particularly vivid for anyone who has witnessed mammalian births as well as the pulsating emergence of the adult cicada from the nymphal cuticle. Should there be any reluctance to recognize a correspondence there, we might consider a passage of Aristotle's work on the natural history of animals (*HA* 556b7) in which, with reference to the emergent cicada nymph, he applies the Greek word *tettigomētra*, which is attested nowhere else in the history of the language. Dictionaries and translations always define the word as "cicada mother," but a more plausible definition, consistent with conventional Greek word formation, might be "cicada womb." The simplex Greek word for womb is *mētra*; that for mother is *mētēr*.

The imagery of the third segment of the excerpt is usually explained by commentators in terms of human birthing and there is certainly a clear analogy drawn there to humans, or at least to mammals, but neither humans nor other mammals correspond to the bulk of the imagery as well as the earth-born cicada does. If the nymphal cicada is a "womb," then, in a sequence that culminates in warm and moist conditions, wombs really do attach themselves to the earth by roots, and in due course detach themselves and instinctively head for the air through perforations that they have made in the moist soil. Then, in their swollen state, open up their integuments to discharge their nascent contents with the aid of lubricating secretions of molting fluid. Is that fluid what prompts Lucretius to describe the emerging adults by the word translated above as "smooth"? While that and other questions remain open, it is hoped that a close reading of the Lucretius' Latin in combination with other documentation and attention to observable natural phenomena described by other ancient writers has belatedly shed some new light on an intriguing piece of entomological poetry.

CHAPTER EIGHT

Insects in Art

GENE KRITSKY

Ancient artists incorporated insects as models for their creations from the very beginning of antiquity over five thousand years ago. Like the insects depicted in carvings and etchings from Prehistory, these first examples were often crude, but through the millennia became intricate high-quality works which inspire artists today.

This change in craftsmanship is easily observed with the honey bee hieroglyph in ancient Egypt. It first appears in several crudely incised carvings from the First and Second Dynasties (Figure 8.1) dating from 3000 to 2687 BCE. Several such examples are on exhibit at the British Museum in London. These small plaques include honey bees that appear more like stick figures than fine works of art, but they do convey the intended meaning. From the earliest times in Egyptian history, the honey bee hieroglyph represented the Delta region of northern Egypt. When combined with the sedge plant and two half circles, which were the hieroglyphs for the consonant "t" you get the ancient phrase "*n-sw-bit*." This translates to "He of the Sedge and the Bee" meaning "King of Upper and Lower Egypt." This royal titulary accompanies nearly every pharaoh's name that is found on temples and tombs walls. This is easily found in hundreds of museums and Egyptian sites by looking for the "cartouche," an oval with a flat base with the hieroglyphs of a pharaoh's name. Cartouches became common in the Fourth Dynasty, just after 2650 BCE. By this time the honey bee glyph had evolved from the stick figure into a carefully rendered insect with the constricted abdomen diagnostic of the insect order Hymenoptera, which includes the honey bees, wasps, and ants (Kritsky 2015).

FIGURE 8.1 An early crude honey bee hieroglyph from William Flinders Petrie (1900). University of Heidelberg Historic Literature Digitized. Public Domain.

Nineteenth-century Egyptologists were not always sure that the honey bee glyph was indeed a honey bee. Sir Wallis Budge illustrated the glyph with the phrase "The Bee or Hornet(?)." Flinders Petrie followed suit in his writings, but by 1923 Petrie had dropped the reference to hornet (Ransome 1986). The confusion is understandable. The hieroglyph has long filamentous antennae which are certainly reminiscent of wasps, and the wings are held upward rather than folded flat over the body. The wasp-like features notwithstanding, the association of the honey bee as part of the word for honey indicates that the ancient Egyptians were associating the insect figure as the producer of honey, and thus it means honey bee (Kritsky 2015).

A hallmark of Egyptian art was its stability, and for nearly three thousand years, except for the brief period when Akhenaten was pharaoh and people

FIGURE 8.2 The cartouche and royal titulary of Ramesses II with the grid used to place the hieroglyphs. Photograph by Gene Kritsky.

were carved in a more realistic manner, the art was relatively unchanged. There was a standard that was followed as to how people and glyphs were carved on tomb and temple walls. Scribes would draw grids on temple and tomb walls to keep the proportions consistent. For example, male figures were drawn on an 18×6 grid with the top of the head tangential to the top of the eighteenth grid, the feet planted on the bottom of the first grid, and the shoulders six grids wide (Robbins 2001).

The cartouche and the royal titulary were also drawn according to a grid pattern with the cartouche with the pharaoh's name in a 4×2 grid and the titulary in a 2×2 grid (Figure 8.2). The ideal honey bee was drawn with the base of the wings at the center of the 2×2 grid. An ideal example would have the hind wing extending back to the upper corner of the grid with the upper middle line bisecting the forewing. The abdomen would extend back to touch the midpoint of the outer line of the lower grid below the hind wing. The scribes could vary this orientation depending on the available room on the wall or column, making a honey bee glyph fit a tighter space by by changing the orientation of the abdomen towards the bottom of the lower hind square and moving the wings closer together if necessary (Kritsky 2015).

Except for the spacing issues, the general style of the honey bee glyph was relatively consistent from the Third Dynasty until the start of the Ptolemaic Period. At the Ptolemaic temple at Kom Ombo there are honey bee reliefs with the wings held in a diverging manner, and the abdomen looking more like the tail of a horse than that of a honey bee. Elsewhere at Kom Ombo temple they are shown with the wings diverging like rabbit ears and the abdomen looking like the scabbard for a scimitar (Figure 8.3) (Kritsky 2015).

FIGURE 8.3 Variations in the honey bee hieroglyph. a) Kalabsha (Ptolemaic Era); b) Kom Ombo (Ptolemaic Era); c) Kom Ombo (Ptolemaic Era); d) Ramsseum (Nineteenth Dynasty). Photographs by Gene Kritsky.

After the bees were carved, they were often painted. At Medinet Habu, the mortuary temple of Ramesses III on the West Bank of Luxor, you can still see some of these painted honey bees. The colors that survive show a blue thorax with the wings and abdomen a reddish ochre that today appears more of a yellow brown.

The honey bee glyph also served as a model for carnelian amulets as seen at the Metropolitan Museum of Art. The Cleveland Art Museum has a Sixth Dynasty button seal with a carved bee on its underside. Bee amulets were also carved out of ivory, serpentine, limestone, cornelian, and as a glazed composition called faience (Andrews 1994; Kritsky 2015).

The most common insect motif in ancient Egypt art was the scarab beetle (Figure 8.4). Some of the carved amulets are detailed to the point that the beetle's species could be identified, *Scarabaeus sacer*. The scarab beetle uses dung for food and to protect its young. These beetles craft two kinds of dung balls, one is a pear-shaped ball made from sheep dung within which the female beetle lays her eggs. When the young hatch, the larvae feed on the dung. These beetles also produce a round ball out of cattle dung, which is only used for food. It was this round ball that captured the theological imagination of the ancient Egyptians as being an earthly metaphor for the scarab beetle god, Khepri, who like the insect, rolls the sun ball across the sky during the day (Wilkinson 2008).

The scarab, like the honey bee, was also a hieroglyph. The ancient word for scarab was *ḫpr*, which is pronounced as "kheper," which was the ancient Egyptian word for "to be created" or the noun for "manifestation." These meanings meant that the beetle hieroglyph would appear in the names of several pharaohs. For example, the great Eighteenth Dynasty Pharoah Tuthmosis III's complete name was Tuthmosis Menkheperre. The scarab also appeared in King Tut's full name, Tutankhamun Nebkheperure. Royal inscriptions were also carefully carved on tomb and temple walls within a grid to be symmetrical and balanced (Clayton 1994).

The scarabs were often painted on tomb walls to represent the god Khepri. In the tomb of the Queen Nefertari, the god is illustrated as a human with a scarab beetle as his head. At Karnak Temple at Thebes, in present-day Luxor, is

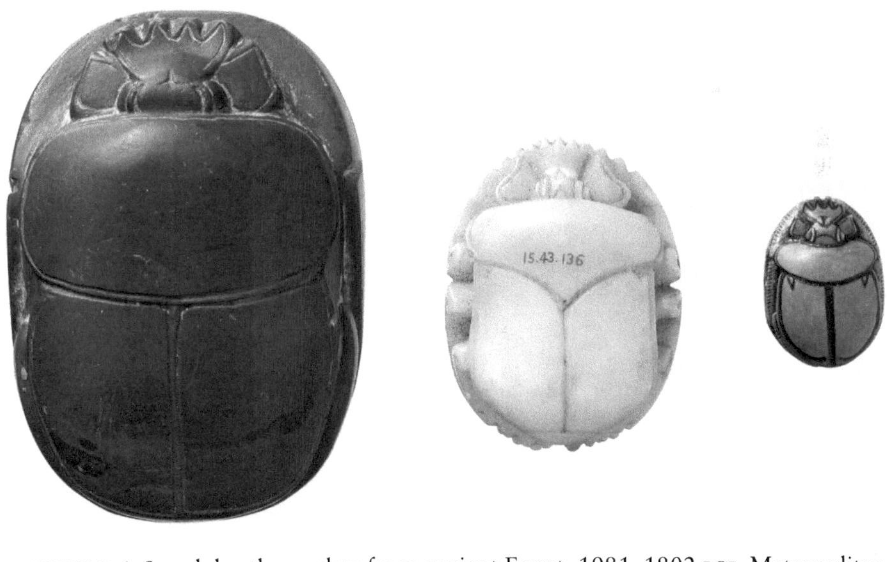

FIGURE 8.4 Scarab beetle amulets from ancient Egypt. 1981–1802 BCE. Metropolitan Museum of Art.

a colossal scarab monument next to the sacred Lake. Another colossal scarab is on exhibit just a few yards from the Rosetta Stone in the British Museum. These large scarab monuments were made of granite and demonstrated the ancient Egyptians' skill at stone working (Wilkinson 2008).

On a smaller scale are the scarabs found on pectorals and other vestments of the pharaoh. These scarabs were made of numerous materials including bronze, carnelian, feldspar, jasper, onyx, lapis lazuli, and glass. One of the most impressive is the scarab on a pectoral found in Tutankhamun's tomb. The large scarab in the center is made of "desert glass" that formed from the heat of a meteorite impact on the sand. A large lapis lazuli scarab mounted on a gold bracelet was also among the treasures of Tutankhamun (Andrews 1994; Wilkinson 2008).

Heart scarabs were relatively flat, smooth, green scarabs which were placed within the wrappings of the mummy. Their purpose was to help the deceased during the weighing of the heart, a ritual judgment that occurred after death to determine if the deceased should proceed to the afterlife. These scarabs were generally between 7 and 11 cm long and usually made from a green stone such as green serpentine, feldspar, or siltstone. Sometimes the underside included spells that would help with the journey through to the afterlife (Wilkinson 2008).

Commemorative scarabs were made during the Eighteenth Dynasty, and as their name suggests, commemorated a major event. For example, a commemorative scarab created during the reign of Thuthmose III referred to the construction of two large obelisks. Thuthmose III successor, Amenhotep III, commissioned commemorative scarabs describing the number of lions and bulls that he hunted, a series that refers to his wife Tiye, the construction of an artificial lake, and the arrival of Princess Gilukhepa, to join the Amenhotep III harem. To date, 190 Amenhotep III commemorative scarabs have been found (Wilkinson 2008).

The most common scarabs found in Ancient Egypt are the smaller scarabs, which included scarab seals and funerary or amulet scarabs. These scarabs were made of a variety of materials including gold, silver, semi-precious stones, steatite, and faience. Steatite, or soapstone, is a soft stone made of magnesium silicate that has a smooth soapy texture which is easily carved (Andrews 1994).

The process of manufacturing scarabs from hard stones was described by Newberry in 1908:

> The process of making a seal out of hard stone was simple enough; a suitable piece of amethyst, jasper, or other material was taken, cut into shape of a ... scarabaeus beetle, and polished. The device or inscription was then engraved in intaglio. In the case of steatite, schist, and other soft stones, the device was

> sometimes drawn in ink before being cut, and the seal was finished by being dipped into a vitreous glaze to harden it ...
>
> The tools used were apparently of four kinds: a knife, a graver, a simple drill, and a tubular drill ...
>
> The knife, perhaps of hardened bronze, was used for cutting the specimens of the softer materials into shape, while the graver, of flint or obsidian, was employed for cutting the device or inscription ....
>
> The simple drill, used for drilling the soft stone seals and for engraving those of hard stone class, consisted of a metal drill with handle, the butt end of which revolved inside a stone or wooden cap which the engraver held in his hand and was thus able to direct the point to the right place. The drill itself was made to revolve by means of rapidly moving forwards and backwards by a bow, the string of which was wound round the stick of the drill ... The tubular drill was also worked in the same way with the bow, but instead of the drill being pointed as in the simple drill, it was tubular. With hard stones both these kinds of drill were used, with emery powder and oil or water.

Faience, more correctly known as glazed composition, was a manufactured material. These artifacts were likely given the name faience because they resembled material made in Faenze, Italy during the Middle Ages (Nicholson 2001).

Egyptian faience amulets could be glazed by three methods. First, the body or core of the amulet was made from crushed silica sand or crushed quartz, materials which were commonly available in Egypt. The crushed silica was mixed with alkaline salts, lime, and copper to form a paste. The paste would be pressed into a clay mold giving the paste the shape of the scarab beetle. This soft casting could have hieroglyphs carved into the base or more detail carved on the surface (Figure 8.5). The cast would then be fired. The firing hardened the core while the alkaline salts and the lime reacted with the core which formed a blue green or turquoise glaze on the surface.

Application glazing involved combining all the ingredients mentioned above to make a powder of the glazing ingredient which was painted on the amulet's core. Objects produced in this manner often showed brush marks or other evidence of application of the glazing material. Heating in the kiln produced the desired turquoise glaze.

Finally, a third process known as cementation was used. This involved placing the core in a glazing mixture that included lime, silica, charcoal, and

FIGURE 8.5 The carved undersides of two faience scarabs. a) honey bee; b) grasshopper. Ancient Egypt, 1479–1458 BCE. Metropolitan Museum of Art.

copper, which served as a colorant. When heated, the surface becomes evenly glazed showing fine details of the object (Nicholson 2001).

Scarab seals were manufactured by the thousands and were often carefully inscribed with hieroglyphs with the owner's name. On the underside would be a phrase or the name and status of the owner. For example, there is a Middle Kingdom scarab that has the phrase Chief beekeeper and King's acquaintance (Kritsky 2015).

These scarabs were used to seal containers. While it would seem insecure to us today, this was a common method of documenting ownership as described by Newberry (1908):

> When the scarab-seal was used for sealing, it was simply pressed upon the clay destined to receive the impression, just as a signet is used at present day. A large number of clay-sealings from scarabs have been found in different localities in Egypt, and bear witness to the manner in which this class of seal was used.
>
> The greater number of scarabs were probably simply strung on a thread of string, by which they were secured to the garment or girdle of the person to whom they belonged. Sometimes they were worn of the finger, attached by a piece of string or they were simply mounted as swivels to metal rings, in which they revolved, or they were enclosed in a metal frame or funda in order to protect the edges from injury, and the mounted as swivels to metal rings …

These scarab seals were different from the funerary scarab amulets, which did not have a phrase inscribed on the underside. Rather, they were true copies of the beetles, with carefully carved legs and a segmented body. Unlike the seals, which were used during the course of the day, these funerary scarabs were attached to the linen wrappings of the mummy to aid in the resurrection of the deceased in the afterlife (Wilkinson 2008) (Figure 8.6).

Scarabs were not the only beetles that attracted the attention of Egyptian artists. Buprestid beetles, also called metallic wood boring beetles, were modeled in faience, gold, and carnelian and strung on necklaces. The most elaborate use was as the head of pins for the royal bed of Queen Hetepheres I, the wife of Pharoah Snefru and the mother of Khufu who build the Great Pyramid. The bed was made of gilded wood and was held together with dowels, the heads of which were gold buprestid beetles (Kritsky 1993).

Possibly the oldest beetle to be found in Egyptian art are elaterid or click beetles. These beetles have an unusual shaped pronotum, a shield-shaped section just behind the head which is a diagnostic feature of the click beetle family, Elateridae. The shape was incorporated in a greywacke composite dish that is in the Petrie Museum Collection at University College, London (Kritsky 1993).

Darkling beetles, belonging to the family Tenebrionidae, were also used as the model for amulets. A fine example is at the Metropolitan Museum of Art in New York, and another can be found at the Petrie Museum in London. These

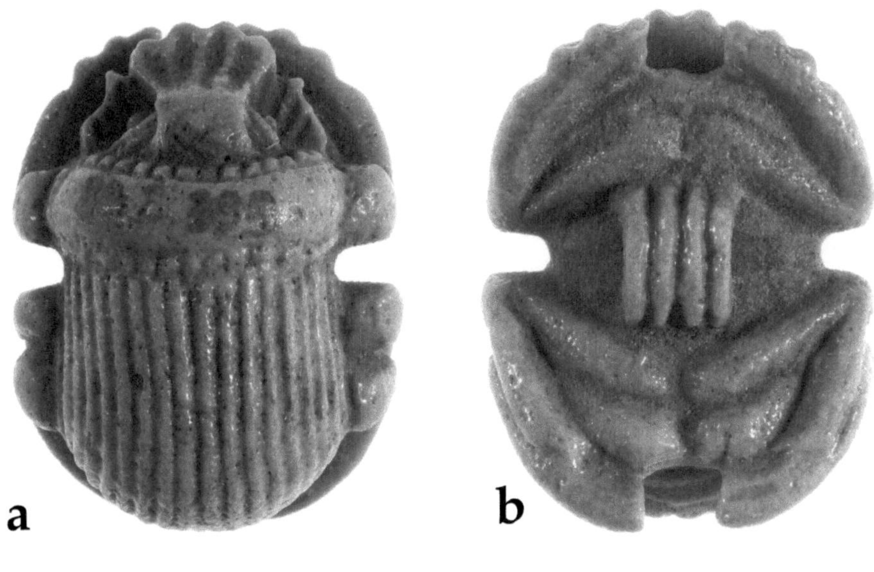

FIGURE 8.6 The upper and lower sides of a funerary scarab, 664–630 BCE. Metropolitan Museum of Art.

rather nondescript beetles are quite distinct from scarabs. Their symbolism likely involves their feigning death when bothered. The beetle tucks its legs close to its body, reminiscent of a mummy, and after a few seconds, the beetle will come back to "life" (Kritsky 1993).

Finally, there is a relatively large beetle pendant in the British Museum which is more reminiscent of a water beetle. It is carved out of a green serpentine stone with incised lines that denote the head, thorax, and two hind wings. A narrower water beetle amulet is found in the Petrie Museum and more representative of a predaceous water beetle belonging to the family Hydrophilidae. These water beetle amulets date to Egypt's First Dynasty.

The Book of the Dead or the Book of Coming Forth by Day, which was produced in the fourth century BCE, includes spells to aid the deceased in the passage to the afterlife. Spell 36 was to repel the beetle or the apshait-insect. Papyri illustrate this spell with a man spearing a generic beetle or scarab beetle, or a man wielding a knife next to an insect that looks like the honey bee glyph with its abdomen, hind legs, and wings removed. Budge (2008) suggested it might be a beetle that feeds on poorly preserved mummies. The clerid beetle, *Necrobia rufipes*, found on several mummies in the past, infests cadavers between twenty to forty days after death to feed on the body (Huchet 2010). Mumification required seventy days to complete, thus this infestation would been noticed by priests during mummification. The identification of *Necrobia rufipes* as the apshait-insect seems obvious when you read the first line of Spell 36: "Depart from me, O thou that has lips which gnaw" (Budge 2008).

Locusts or grasshoppers were not as common in ancient Egyptian art as honey bees and scarabs, but they are also found on tomb walls, used as a hieroglyph, and as amulets. At the late Fifth Dynasty tomb of Kagamni and the Sixth Dynasty mastaba tomb of Mereruka at Saqqara are bas-relief scenes of life along the Nile, including locusts among the papyrus plants. The insects depicted clearly have jumping hindlegs of locusts, but the examples in flight have narrow wings found on dragonflies. In either case, these reliefs were carved by individuals not intimately familiar with grasshopper or dragonflies (Kritsky 1993).

Locusts were also used as a hieroglyph with a beautiful example found at Ramesses III's mortuary temple, Medinet Habu. Locust hieroglyphs could be interpreted as "locust," but at Medinet Habu it was used to denote "numerous soldiers."

Locusts were found as amulets made of steatite and faience during the Old Kingdom to the Eighteenth Dynasty. Andrews (1994) suggests that due to their ability to reproduce in numbers they may have provided fertility to the wearer. There are two utterances in the Pyramid texts that state the deceased ascended to heaven in the form of a grasshopper. At the Metropolitan Museum of Art are faience amulets that resemble to flying locusts from the tomb of Kagamni, but without the jumping hind legs making them appear as stylized dragonflies. These amulets date to the Middle Kingdom's Twelfth and Thirteenth Dynasties.

Butterflies are colorful insects that did not escape the attention of the ancient Egyptian artists. They are often found in tomb paintings from the Old Kingdom through the New Kingdom to illustrate the afterlife, a time of fishing and fowling in a beautiful marsh with plenty of fish, birds, and butterflies. One of the most striking examples is from the tomb of Nebamun and likely from the Valley of the Nobles near the Valley of the Kings on the west side of the Nile in present day Luxor. The exact location of the tomb is unknown as the paintings were all cut away from the tomb walls and now adorn the walls of second floor of the Ancient Egypt hall of the British Museum in London. The marsh scene shows Nebamun standing on a papyrus boat with his wife and daughter. He is holding a club in his left hand and holding onto the legs of a bird in his right. There are fish in the river, papyrus growing along the banks with a cat standing on a papyrus plant biting the wing of a bird. Seven butterflies are flying about in the scene that is thought to be an idyllic representation of the afterlife. The butterflies are carefully painted and can be identified to the species *Danaus chrysippus*.

There is a similar scene in the Eighteenth Dynasty tomb of Nahkt. This marsh has nesting birds and three butterflies (one is partially destroyed) and a possible "dragonfly" figure. These insects are not as clearly rendered as Nebamun's tomb. However, it is likely that the darker butterfly with the white spots may be a male *Hypolimnas misippus*. The tomb of Menna, which is near to Nakht's tomb, also has a marsh scene. Menna'a tomb has been carefully restored, and the paintings are bright and detailed. The artist took more liberties with the butterflies than did the artist of Nebamun's butterflies (Figure 8.7).

Whereas some of these painted scenes are detailed to permit species identification, that is not always the norm. Some butterflies are crudely painted and cannot be identified with any degree of certainty; however, there are some butterflies that are accurate enough for identification. These identifications have been valuable for historical biogeography purposes.

Stylized butterflies were also incorporated into jewelry. The tomb of the mother of Khufu, the builder of the Great Pyramid, was found to have fourteen silver bracelets, inlaid with semiprecious stones arranged in the shape of a butterfly. The Metropolitan Museum of Art's collection includes some rather stylized faience butterfly amulets that date back to the Twelfth–Thirteenth Dynasties of the Middle Kingdom.

The Egyptian depiction of flies is greatly stylized compared to some of the butterflies. The more common examples are small amulets that adorned the necklaces of mummies. These flies were made of gold, silver, carnelian,

FIGURE 8.7 Butterflies in a marsh scene in the tomb of Menna. Photograph by Gene Kritsky.

steatite (soapstone), and faience, and their construction was determined by the materials used. The silver and gold flies were likely cast, whereas those made of semiprecious stones and ivory were carved (Figure 8.8).

Flies were also made into large pendants that were presented as a military or some kind of royal award. The most famous of these flies are three large golden flies. The provenance of the flies has recently been reviewed by Marshall (2015). The flies were made by repoussé of the head and the thorax, which was then soldered to a gold sheet cut to the shape of the wings. The head includes two large compound eyes followed by a thorax made by cutting five tapering rectangles in gold that converged at the end of the thorax made of two transverse cuts in the gold. They were found on a woven gold necklace. These flies are quite different compared to other large fly pendants which were made of mixed materials including ivory and gold and electrum (an alloy of silver and gold). The flat wings were carved out of ivory with a simple head with eyes that are only suggested with incised lines at the tip of the pendant. The thorax is similarly simple with only an incised lines. In the Petrie Museum in London

FIGURE 8.8 An ivory carved fly amulet and a small glazed carved steatite fly amulet, 1550–1525 BCE. Metropolitan Museum of Art.

are two flies carved from bone. Their style is reminiscent of the ivory and metal flies, except they lack any metal. The head and thorax formed in high relief of bone. In these flies the eyes are depicted as two elongated ovals.

Fly pendants were also carved on statues and painted on human figures. An example of the former is on display in the Cairo Museum. It was found at Edfu Temple and is of a kneeling man wearing large fly pendants on a necklace not unlike the golden flies. A painted example is found in the Eithteenth Dynasty tomb of the Vizier of Rekhmire in the Valley of the Nobles on West Bank at Luxor. The scene shows two Nubians, one leading a giraffe and another man carrying a tribute. The man leading the giraffe is wearing a necklace festooned with a single large fly, whereas the tribute bearer's necklace has three large flies (Marshall 2015).

Although not an insect, there is another arthropod that features in ancient Egyptian art, the scorpion. Scorpions were poisonous animals that inflicted their painful stings on the Egyptians to the level that there were magical spells recorded to repel the creatures. "Scorpion" was a predynastic king of Egypt before the unification. Evidence of his existence are two mace heads in the collection of the Ashmolean Museum in Oxford depicting the king with a carefully carved scorpion front of his face. Those images are used as the king's identification.

Scorpions were more common on vessels in Egypt's Old Kingdom but not during later periods. Amulets in the form of scorpions were possibly carried to act as a talisman to repel the pest.

A stylized scorpion adorns the head of the goddess Selket, one of the four goddesses that protected the shrine that housed the canopic chest found in the tomb of Tutankhamun. This wooden gilded statue holds her arms wide and is looking to the left. Upon her head is a scorpion. The statue created quite a stir when it toured the United States in 1976.

The review of the insects that the Egyptians were illustrating in their papyri, reliefs, carvings, amulet, jewellery, and the like does not suggest that they were focusing on minute details. Some scarabs were produced with such detail that is clear what species was represented. However, there are thousands of examples which cannot be taken to species but simply as a scarab. The various illustrations of the "beetle" in the Book of the Dead suggest that the scarab hieroglyph was used interchangeably as beetle. This is similar to the confusion over the hieroglyph for honey bee as well.

As you read in Chapter One by Ian Beavis, our knowledge of insect diversity expanded with the ancient Greeks and Romans. In addition to their written works, we can consult their art for details of the insects that enticed their imaginations, and we will do so by examining the gemstones with carved insects. These engraved gems, also called intaglios, were small wearable art, which have provided insight into the ancient world's knowledge of insects. The skill required to create an intaglio was outlined by Henry Middleton (1891) (Figure 8.9).

> The tools used by the ancient gem-engravers were mainly tools for of three different sorts, namely the *drill*, the *wheel* and the "diamond-point." In using all these the gem itself was firmly fixed in a bed of cement made of pitch and pounded pottery. The tools were held in the engraver's hand, who thus had a greater freedom of touch than a modern gem-engraver, who usually works in the reverse way, having his drills and wheels fixed, and the gem loose in his hand.
>
> THE DRILL The drill was worked in the old fashion, which still survives in the East: the string of a small bow was wound round the stick of the

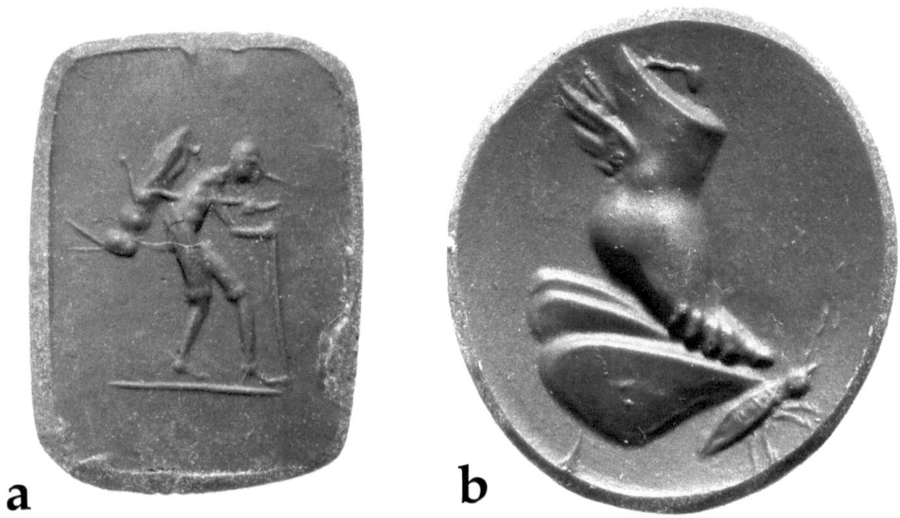

FIGURE 8.9 Impressions from two intaglios, both first century BCE–third century CE. a) from a carved sard showing a man carrying a locust; b) a carved garnet showing a winged foot on a butterfly. Metropolitan Museum of Art Collection.

drill, which was made to revolve by moving the bow rapidly backwards and forwards. The butt end of the drill revolved inside a cap or tube, which the engraver held in his hand, and so directed the point to the right place; he could not, of course, hold the drill by its revolving part.

It is only long practice that enables a workman to manage successfully the difficult task of carrying on simultaneously a distinct movement with each hand; so in some cases, especially when working larger sculpture, an assistant worked the bow, leaving the engraver free to direct the point of the drill ...

In gem-engraving the point of the drill, which cut into the stone, was of soft metal, usually bronze, and varied in size from that of an ordinary pin to a good-sized knitting needle or even larger. The actual cutting of the drill was done, not by the metal, but by the fine emery powder which, mixed with oil, was kept constantly smeared upon it. The minute particles of emery, which is a form of *corundum*, stick in the soft metal under the pressure, and so give a steady cutting surface ...

In most of the archaic gems, and again in those of the period of Roman decadence, the use of the drill is very conspicuous; short curly hair of men or animals is often represented by a series of close-set drill holes even in work of a good period.

> The general blocking out of the figures seems to have been mainly done with the bow-drill, the final modelling and details being put in afterwards with other tools ...
>
> THE WHEEL: this was a minute disc of bronze which was set on a long, slender shaft of wood or metal and worked with a bow and tube like the drill; emery and oil being applied to it in the same way. The wheel cut, of course, at right angles to the shaft, not in the same direction as the drill did. According to the direction in which the workman moved the little wheel it could cut either a long line or a broad sunk surface.
>
> In the early lenticular gems its use is specially visible, and indeed to use the wheel for any other purpose than blocking out the design requires exceptional skill on the part of the operator ....
>
> THE DIAMOND-POINT: this tool was not used with the bow, but was held in the hand like a pencil, or like the "dry-point" of a modern etcher.
>
> THE FINAL POLISH. After the sunk design of an engraved gem was completed, it was necessary, both for the sake of its beauty, and also to prevent the wax or clay of the seal from adhering to it, to polish, as completely as possible, the internal sunk part.

When completed, the intaglios were mounted in gold rings, and, when pressed into clay or wax, would produce the engraved image as a raised relief.

A survey of insects carved on intaglios includes the insects found on ancient Egyptian art and expands on that diversity. A search of the intaglios that include the word "insect" in the collections catalog at the British Museum in London produced 186 entries. The most common insects in the collection are butterflies. These include a banded agate with an intaglio of two butterflies flanking a plant, and an orange chalcedony intaglio with a butterfly standing or supporting scales. The Museum's collection includes several ant intaglios, the ants being recognized by their constricted abdomen and wingless condition. Among the intaglios of flies is a very detailed fine example of a cornelian engraved with a fly surrounded by a wreath of olive branches. Another cornelian gemstone has a grasshopper engraved in profile. Finally, there is a rather humorous engraving of a grasshopper walking upright on its jumping legs and carrying two stalks of wheat.

Two intaglios illustrated in other collections are noteworthy. In the Hermitage is a fine wasp engraved in rock crystal. The detail is enough to permit its family identification. An intaglio in a private collection that dates from Greece's first century BCE illustrates the life cycle of a butterfly with the pupa above the adult winged butterfly, and the caterpillar stage below the adult (Davies et al. 1986).

Many of the same insects found on gems were also found on ancient coins (see Figure 5.1), and several of these can be viewed at the British Museum. The process of making coins is closely aligned with that of making intaglios, except rather than starting with a gemstone, the coin starts with a blank disk of gold, silver, or an alloy of the two called electrum. The blank was often cast from a mold into which molten metal was poured. Once cooled, the blank could be removed from the mold, and any excess metal cut away. The blank was then placed between two dies and struck with a heavy hammer conferring the reverse image from the die on each side of the coin. Dies were made in a process similar to engraving an intaglio, as described by Hill (1922):

> The method of engraving the designs on the dies must be inferred from extant remains; for there are no literary records of it. It is generally assumed that the arts of die-engraving and gem-engraving were in all essentials the same; both were directed to producing a sunk design in a hard material ... It has been generally agreed that the gem-engraver's principal tool, the drill (driven by wheel or bow), was also used by the die-engraver, although it has also been generally admitted that he used a cutting tool, such as a burin or graver, as well .... It is probable that the work was done chiefly with punches and scorpers.

The obverse side, or the principal side of the coin, could be struck up to fifteen thousand times depending on the metal of the die and the blank with the reverse wearing down with half as many strikes. The obverse was often carved deeper and with more detail than the reverse. It was placed in the anvil, which further protected it.

The symbol of Ephesus was the honey bee, and the Ephesians included a large honey bee on the obverse of their coinage starting around 400 BCE (see Figure 5.1). Grasshoppers, or locusts, were struck on early coins, and as with the intaglios, the locust was sometimes depicted with a head of wheat. A fifth-century BCE coin has a praying mantis in the same pose as the locust to the right side of a wheat head. Another coin includes a grasshopper on the edge of a coin with two eagles killing a hare. Small bees, and sometimes cicadas, were often included on the reverse of ancient coins. The determination between a bee or a cicada is based on the orientation of the wings and the eyes. Bees have the wings spread out from the body, whereas the cicadas have wings held parallel to the body and have larger eyes. Another arthropod common on ancient coins is the scorpion. Small scorpions may be seen on the obverse side of a 90 BCE Roman coin behind the head of Apollo (Forum Ancient Coins 2022).

Much of what we know about life in ancient Greece comes from the study of Greek pottery. When archeological sites were plundered, the thieves were looking for gems and precious metals. Fortunately, for our purposes these looters were not interested in pottery with beautifully rendered scenes of life during ancient Greek history. Huddliston (1902) wrote of this pottery, "The graceful outlines that the potter's wheel produced are inimitable; the rich black and red lustre, that seems still to shine as brightly as it could have done when turned out of the workshop, defies reproduction."

To better understand the mindset that would produce such works of art, it is important to know the process by which this pottery was made. A process that required a knowledge of clay and how to prepare it. This is an ancient knowledge as Richter described in 1924:

> The chief ingredients of clay are silica, alumina, and water. Other possible ingredients are iron oxide, lime (calcium oxide), magnesia, and potash. To the iron compounds are due the different colors of the clay. When potters speak of the color of a clay—red, yellow, white—they refer to the color after burning, not in the raw state. The tones of the color are controlled by heat; for instance, a red clay becomes first pink, then in a higher fire a deeper red, and in a still higher fire a brownish red ...
>
> The potter demands three properties of his clay: (1) plasticity, the property which enables the clay to acquire form; (2) porosity, the property which enables the water to escape; and (3) vitrification, the property which enables the clay to be fired ...
>
> When the right composition of the clay has been assured, the next step is to wash it and separate it from the many natural impurities, such as stones, sticks, etc ... But even at this stage the clay is not yet ready for use; it has first to go through the important process of kneading or "wedging." This consists of cutting a ball of clay in two against a wire, slapping the two parts on a plaster or wooden board, one on top of the other, then lifting up the whole lump, cutting it in two again, and slapping it down as before. The purpose is to remove all air bubbles and to correct irregularities in hardness. The operation has sometimes to be repeated fifteen or twenty times before a good texture is secured.
>
> The first task in fashioning a vase on the wheel—or "throwing" it, as it is technically called—is to center the ball of clay on the wheel-head. It is accomplished by pressing the left hand against the ball of clay as it revolves rapidly, care being taken to keep the left forearm absolutely rigid. The right hand is used for keeping the clay wet by sprinkling it with water, and for

> pulling the clay inward, thus squeezing it up to a cone ... When the [cone] runs perfectly true, it is time to open it by inserting the thumb in the center; then by placing the fingers of one hand inside the hollow, and the fingers of the other on the outside, and by raising both hands gradually [outward] several times and squeezing the clay lightly while so doing, a cylinder is formed.

After the general shape of the vase is formed, it was allowed to dry for a day and then any excess clay could be carefully trimmed away, and the surface smoothed using a damp sponge or cloth. After this step, a foot (the flared base of the vase) and curved handles could be added. Both the foot and handles were also made of clay. At this point the vase could be prepared for painting. First any areas of the vase that were to remain black were painted with slip, a fine clay. When the vase is fired these slip-painted regions would be black. Areas where a scene was to appear were left unpainted. The vase was then allowed to dry to a harder leathery texture.

The potter could then make a charcoal sketch of the desired scene on the unpainted region of the vase. Slip would be painted on areas that were to be black. After the slip had dried, it could be incised to produce more fine detail in the scene. Red ocher could be painted on regions of clothing or other details. When fired, these would appear purple. White regions could be produced by painting clay with little iron oxide on the desired sections.

The firing process required considerable skill and experience given the nature of kilns and the lack of thermometers. The firing occurred in three stages. When the kiln was approximately 1500° F the vase would be placed in the kiln. Air was allowed to enter the kiln, and the oxygen caused the clay and slip to turn red. The kiln was made hotter, the air was stopped, and damp wood or plants were placed in the kiln, which resulted in the clay and slip turning black. It also gave the slip a glossy appearance and a hard surface. Finally, air was permitted to flow through the kiln again, which resulted in the unpainted surfaces to turn red as the slip-covered sections remained black. The vase would be removed from the kiln after it had cooled for at least half of a day.

Several insects have been illustrated on Greek pottery including bees, wasps, locusts, and cicadas. In some instances, these insects are generalized to the point where they can be identified to a particular order, and if there is more detail in the illustration to family, and rarely species. In the Boston Museum of Fine Arts is a libation bowl that has a cicada not drawn as part of the illustration but rather a three-dimensional cicada mounted at the center of the bowl. The detail in the cicada is sufficient to assign it to family and even genus.

The insects and arthropods depicted on Greek pottery can in general provide clarification both for the art and the biology inferred. This is best seen on a vase in the collection in the British Museum. This particular vase was purchased in

FIGURE 8.10 A terracotta amphora showing naked soldiers stealing Zeus' honey being attacked by wasps. Pottery: black-figured amphora. © The Trustees of the British Museum.

1847 from an Italian collector known as Basseggio (Figure 8.10). The catalog description of the front of the vast in 1851 reads:

> Clay pale, varnish black 5 design black, white and crimson, with incised lines; 1. on a tablet, four male figures stung by bees, probably the four who plundered the hives from the honey of which Zeus was nourished when an infant; their names were Laios, Keleos, Kerberos and Aigélios; they are represented entirely naked; the first on the right is crouching down, holding in his hands the circular disk-like stone, on which one of the hives had been placed; single bees are stinging him on the head, breast, left knee and penis; next to him is a figure throwing up his arms and endeavoring to beat off the bees with a branch; he is stung on the breast and left arm; a bee is flying at his head, and another at his right side; the third figure is squatting down, his face to the front; he raises his left hand towards his head which a bee is stinging, another stings him on the right shoulder, a third on the penis, which is inflamed with the poison; and a fourth on the left foot; a fifth bee flies at his left knee; the last figure brandishes a branch in his right hand, raising his left foot in agony, and with his left hand brushing off a bee which

has stung him on the left knee; he is also stung on the shoulder, back, and right foot; a bee is flying at his right knee, another at his left foot and a third at his right elbow; all these figures are bearded, the last in the group wears a diadem.

(British Museum 1851)

The Zeus myth, thought to have inspired this scene, involves a cave where the infant Zeus has been hidden along with beehives to provide nourishment. The men, in an attempt to steal the honey for themselves, put on their armor and entered the cave. When they saw Zeus in his cradle, the joints of their armor broke and it fell off. Once naked, they were attacked by the insects (Ransome 1986).

There is a second amphora in Basel at the Antikenmuseum which shows a similar scene. However, the bee scene includes only three men instead of four. The man on the far right is holding a disk-like structure, while the other two are being horribly stung as they try to ward off the bees with branches.

The differences in personnel notwithstanding, there are details which are in the illustration that were not considered in many of the descriptions. First, the stone disk that one of the men is holding is not the base of a hive, but more likely the front cover of a horizontal hive. Several horizontal clay hives dating to 400 BCE have been found in Greece, and such hives were equipped with a disk-shaped cover which was held in place with rope. I have also visited traditional horizontal hives on Syros and Paros that use similar disks placed at the entrance of horizontal hives and held in place by stones placed in front of the cover. Bees would enter the hive by gaps formed by the lid not being tightly fitted to the hive.

Another, and more obvious, problem involves the “bees.” In both amphorae, the “bees” are clearly not honey bees, but rather hornets. Unlike the Basel amphora, the British Museum amphora’s insects are carefully illustrated, and they include the lengthwise folded wings and the diagnostic abdominal markings of the Oriental hornets. The Oriental hornets are major robbers of honey bees, and I have observed these hornets’ aggressive behavior against beehives while visiting a traditional apiary in Paros. This identification of the hornets as bees is not unusual as even today most people do not get close enough to tell the difference. What is the more likely scenario is that the hornets were trying to rob the beehives, when the men arrived, and it was the hornets attacking the beehives attempting to stop the humans from robbing the hives that the hornets were pillaging.

Many of the examples of insects in art during antiquity have dealt with tombs, amulets, or jewelry. However, we do have evidence of insects in art that is more aesthetic than we have observed in the past, and these examples are found in the artifacts from Pompeii. Pompeii was a Roman city that was destroyed in 79 BCE from the ashfall from the eruption of Mount Vesuvius. The city was stopped in its tracks for centuries before excavations provided a vivid picture of life at the time of the destruction of Pompeii. The artworks that have been preserved are in the form of carvings, mosaics, and paintings. While they do not document significant insect diversity, they do illustrate that insects were part of natural world as perceived by these ancient artists.

An important example is found with the carved marble portico of the entrance to the Eumachia building, which was east of the Forum. This portico included ornate carvings of acanthus leaves with birds and insects. The insects are detailed carvings, but they can be identified to orders, and in a few cases to families. The insects included are stylized butterflies, ants, beetles, flies, and grasshoppers. There are six-legged spiders among the reliefs. Of these carvings, the grasshoppers are the best rendered as they are easily identified by their jumping hindlegs (Larew 2002).

Other insects can be found on mosaics and some frescos. The mosaics are, as described by Anthony (1935), "a surface decoration made by setting small pieces of differently colored stone or glass in a bed of cement." These "pavements" were durable and survived the eruption providing us with a glimpse of these decorative artforms. The insects depicted in mosaics include grasshoppers and butterflies, the latter of which were mostly elements of butterflies as part of a larger scene. A good example was found on a tabletop in the tannery garden. It shows the narrow wings of a moth between a human skull and a wheel. It is believed that the wings represented the soul or psyche. Grasshoppers or possibly crickets are found in mosaics in the house of Doves and on a basin mosaic in the house of Menander.

Insects also appear on a few frescos preserved in the ruins of Pompeii. Roman frescos required special preparations of the surface to be painted, as was described by Merrifield (1846):

> We now come to the execution of fresco paintings. Of all kinds of painting, that of fresco painting is the most masterly, dexterous, and expeditious, completing in one day, what, in any other style, would occupy a long time in painting, and which may be retouched. But fresco painting requires great dexterity and boldness, and the mistakes and faults are irremediable, unless the work be cut out. It is the most manly and lasting kind of painting, therefore, those who practice it ought to be more respected and esteemed than other painters, being considered great masters.
>
> The wall must be very dry, strong, and free from moisture, and must have been plastered a long time previous; and the lime which is to be used must be very mild, it being necessary to keep it in soft water above two years, and it must be mixed with fine sand, in equal proportions, and the mortar must be laid on that part only which can be painted in one day.

Some insects can be found on the frescos at Pompeii. A butterfly can be seen in a garden painting in the house of Adonis, and two butterflies adorn paintings in the peristyle of the Villa of Diomedes. A very stylized painting of a butterfly in the Villa of Poppea is similar to one of the carvings in the marble portico of the Eumachia building. A clearly rendered grasshopper also adorns the walls of the Villa Poppea (Larew 2002).

One of the more fanciful insect illustrations in Herculaneum is a drawing of a parakeet pulling a chariot driven by a cricket (Figure 8.11). This is reminiscent of a carved intaglio with a cricket driving a chariot that is pulled by two dogs (Larew 2002).

Comparing the insects depicted in ancient art with the insect knowledge during antiquity shows a disparity in insect diversity. As the diversity of insects mention in the ancient literature became more robust with the writings of the ancient Greeks and Romans, that diversity was generally not matched with increasing insect diversity as we approached the Common Era with bees, beetles, butterflies, grasshoppers, flies, and cicadas the most common insects catching the eyes of artists during antiquity.

FIGURE 8.11 A cricket driving a chariot pulled by a parrot. From Le antichità di Ercolano esposte. Napoli, nella Regia Stamperia (1757), Archive.org.

BIBLIOGRAPHY

Adrados, F.R. 1999. *History of the Graeco-Latin Fable*, vol. 1: *Introduction and From the Origins to the Hellenistic Age*. Trans. N.D. Ray. Leiden: Brill.

Aldred, C. 1971. *Jewels of the Pharaohs*. New York: Praeger Publishers.

Alfieri, A. 1931. "Les insectes de Ia tombe de Toutankhamon." *Bulletin de la Société Royale Entomologique d' Egypte* 31: 188–9.

Alluaud, C. 1908. "Notes sur les coléoptères trouvés dans les momies d' Égypte *Bulletin de la Société Royale Entomologique d' Egypte* 1: 29–36.

Andres, A. 1931. "Catalogue of the Egyptian Tenebrionidae." *Bulletin de la Société Royale Entomologique d' Egypte* 15: 74–125.

Andrews, C. 1994. *Amulets of Ancient Egypt*. London: British Museum Press.

Andrews, C.A.R. 2003. "Amulet." In D.B. Redford (ed.), *The Oxford Essential Guide to Egyptian Mythology*. New York: Berkley Books.

Angel, G.K. 1966. "Porotic Hyperostosis, Anemias, Malarias, and Marshes in the Prehistoric Eastern Mediterranean." *Science* 153: 760–3.

Antelme, R.S. and Rossini, S. 2001. *Sacred Sexuality in Ancient Egypt*. Trans. J. Graham. Rochester, VT: Inner Traditions International.

Anthony, E.W. 1935. *A History of Mosaics*. Boston, MA: Porter Sargent.

Anton, K.-W., Halperin, J., and Calderon, M. 1997. "An Annotated List of the Bruchidae (Coleoptera) of Israel and Adjacent Areas." *Israel Journal of Entomology* 31: 59–96.

Apicius. 2009. "Cooking and Dining in Imperial Rome." Available online: http://www.gutenberg.org/files/29728/29728-h/29728-h.htm (accessed April 19, 2023).

Aristophanes. 1853. *Clouds*. Ed. W.J. Hickie. Available online: http://www.perseus.tufts.edu/hopper/text?doc=Perseus:abo:tlg,0019,003:723 (accessed April 19, 2023).

Aristotle. 1910. *Works of Aristotle: Historia Animalium*. Vol. 4. Ed. D.W. Thompson. Oxford: Clarendon Press.

Arnold, D. 1999. "The Middle Kingdom and Changing Notions of Kingship." In F. Tiradritti (ed.), *Egyptian Treasures From the Egyptian Museum in Cairo*. New York: Harry Abrams, Inc.

Arrian (Flavius Arrianus Xenophon). 1986. *The Campaigns of Alexander*. Trans. A. de Sélincourt. London: Penguin Books.
Attia, R. and Kamel, A.H. 1965. "The Fauna of Stored Products in U.A.R." *Bulletin de la Société Royale Entomologique d'Egypte* 49: 221–32.
Atty, D.B. 1983. *Coleoptera of Gloucestershire*. Cheltenham.
Bancroft-Hunt, N. 1979. *People of the Totem*. London: Orbis Publishing.
Barkai, O., Lernau, O., and Kahanov, Y. 2013. "Analysis of Fish Bones From the Tantura F Shipwreck, Israel." *Archaeofauna* 22: 189–99.
Barker, F. 1972. "*How They Found the King*" (compiled to mark the exhibition Treasures from the Tomb of Tutankhamun, at the British Museum, from March 29, 1972). London: New English Library Ltd.
Beavis, I.C. 1988. *Insects and Other Invertebrates in Classical Antiquity*. Exeter: University of Exeter.
Bégouen, M.M. and Bégouen, L. 1928. "Découvertes Nouvelles dans la Caverne des trois frés a Montesquieu-Avantés (Ariége)." *Revue Anthropologique* 38: 358–64.
Bellés, X. and Halstead, D. 1985. "Identification and Geographical Distribution of *Gibbium aequinoctiale* Boieldieu & *G. psylloides* (Czenpinski) (Coleoptera: Ptinidae)." *Journal of Stored Product Research* 21: 151–5.
Berenbaum, M. 1995. *Bugs in the System*. New York: Addison-Wesley Publishing Co.
Berenbaum, M. 2010. *Honey, I'm homemade: Sweet Treats from the Beehive Across the Centuries and Around the World*. Urbana: University of Illinois Press.
Berrens, D. 2015. "Bienen und Literatur: Überlegungen zu Senecas 84. Lucilius-Brief." *Antike Naturwissenschaft und ihre Rezeption* 25: 145–64.
Berrens, D. 2018. *Soziale Insekten in der Antike: ein Beitrag zu Naturkonzepten in der griechisch-römischen Kultur*. Göttingen: Vandenhoeck & Ruprecht.
Beutelspacher, C. 1989. *Las mariposas entre los antiquos mexicanos*. Mexico: Fonda de Cultura Economica.
Bianucci, R., Mattutino, G., Lallo, R., and Charlier, P. 2008. "Immunological Evidence of *Plasmodium falciparum* Infection in an Egyptian Child Mummy From the Early Dynastic Period." *Journal of Archaeological Science* 35: 1880–5.
Bierlein, J. 1994. *Parallel Myths*. New York: Ballentine Books.
Birdsong, R.E. 1934. "Insects of the Bible." *Bulletin of the Brooklyn Entomological Society* 29: 102–6.
Black, J. and Green, A. 1992. *Gods, Demons and Symbols of Ancient Mesopotamia: An Illustrated Dictionary*. London: The British Museum Press.
Bloch, G., Francoy, T.M., Wachtel, I., Panitz-Cohen, N., Fuchs, S., and Mazar, A. 2010. "Industrial Apiculture in the Jordan Valley During Biblical Times with Anatolian Honeybees." *Proceedings of the National Academy of Science* 107(25): 11240–4.
Bodenheimer, F.S. 1951. *Insects as Human Food: A Chapter of the Ecology of Man*. The Hague, Netherlands: Dr. W. Junk.
Bonet, R.H. and Mata Parenõ, C. 1997. "The Archaeology of Beekeeping in Pre-Roman Iberia." *Journal of Mediterranean Archaeology* 10: 33–47.
Borror, D.J., Triplehorn, C.A., and Johnson, N.F. 1992. *An Introduction to the Study of Insects*, 6th edn. New York: Saunders College.
Bostock, J. 1855. *Pliny the Elder the Natural History. Book IV* (ch. 15). Available online: http://data.perseus.org/citations/urn:cts:latinLit:phi0978.phi001.perseus-eng1:11.15 (accessed April 19, 2023).

Boyer, R.S., Rodin, E.A., Grey, T.C., and Connolly, R.C. 2003. "The Skull and Cervical Spine Radiographs of Tutankhamun: A Critical Appraisal." *American Journal of Neuroradiology* 24: 1142–7.

Bozzoli de Wille, M.E. 1975. "Birth and Death in the Belief System of the Bribri Indians of Costa Rica." PhD thesis, University of Athens, Athens, GA.

Braudy, L. 2016. *Haunted: On Ghosts, Witches, Vampires, Zombies, and Other Monsters of the Natural and Supernatural Worlds*. New Haven, CT: Yale University Press.

Bray, R.S. and Anderson, M.J. 1979. "*Falciparum* Malaria and Pregnancy." *Transactions of the Royal Society of Tropical Medicine and Hygiene* 73: 427–31.

Breasted, J.H. 1930. *The Edwin Smith Surgical Papyrus*. Vol. I. Chicago: The University of Chicago Oriental Institute Publications III. Reissued by The Oriental Institute, 1991.

Brendell, M.J.D. 1975. "*Coleoptera: Tenebrionidae. Handbooks for the Identification of British Insects, V, 10*." Royal Entomological Society of London.

Brill, T.B. 1980. *Light: Its Interaction with Art and Antiquities*. Berlin: Springer.

British Museum. Department of Greek and Roman Antiquities, Newton, C.T. and Birch, S. 1851. *A Catalogue of the Greek and Etruscan Vases in the British Museum*. London: Printed by W. Nicol.

Brookesmith, P. 1997. *Future Plagues* (Biohazard, Disease and Pestilence). London: Blandford.

Brothwell, D.R. 1991. "On Zoonoses and Their Relevance to Paleopathology." In D.J. Ortner and A.C. Aufderheide (eds.), *Human Paleopathology*, 18–22. Washington and London: Smithsonian Institution Press.

Bruce, W.G. 1958. "Bible References to Insects and Other Arthropods." *Bulletin of the Entomological Society of America* 4: 75–8.

Bruce, W.G. 1992. "Bible References to Insects and Other Arthropods." In J. Adams (ed.), *Insect Potpourri: Adventures in Entomology*. Gainesville, FL: Sandhill Crane Press.

Bryan, C.P. 1930. *Ancient Egyptian Medicine: The Papyrus Ebers* (trans. from the German). Chicago: Ares Publishers Inc.

Buck, F.D. 1958. "The British Anobiidae." *Proceedings of the South London Entomological & Natural History Society*: 51–64.

Buckland, P.C. 1982. "The Malton Burnt Grain: A Cautionary Tale." *Yorkshire Archaeological Journal* 54: 53–61.

Bucklow, S. 2016. *Red: The Art and Science of a Colour*. London. Reaktion Books.

Budge, E.W.B. 1988. *From Fetish to God in Ancient Egypt*. New York: Courier Dover.

Budge, E.W.B. 2008. *The Egyptian Book of the Dead*. London: Penguin Books.

Buikstra, J.E., Baker, B.J., and Cook, D.C. 1993, "What Diseases Plagued Ancient Egyptians? A Century of Controversy Considered." In W.V. Davies and R. Walker (eds.), *Biological Anthropology and the Study of Ancient Egypt*, 24–53. London: British Museum Press.

Bulfinch, T. 1991. *Bulfinch's Mythology*. New York: Harper Collins.

Bunson, M. 1991. *The Encyclopedia of Ancient Egypt*. New York: Facts on File.

Burfield, T. and Reekie, S-L. 2005. "Mosquitoes, Malaria and Essential Oils." *The International Journal of Aromatherapy* 15: 30–41.

Burgess, N.R.H. and Cowan, G.O. 1993. *A Colour Atlas of Medical Entomology*. London: Chapman and Hall Medical.

Cambefort, Y. 1987. "Le scarabée dans l'Égypte ancienne. Origine et signification du symbole." *Revue de l'histoire des religions* 204(1): 3–46.

Cambefort, Y. and Hanski, I. 1991. "Dung Beetle Population Biology." In I. Hanski and Y. Cambefort (eds.), *Dung Beetle Ecology*, 36–50. New Jersey: Princeton University Press.

Campbell, G. 2003. *Lucretius on Creation and Evolution: A Commentary on* de Rerum Natura. Oxford: Oxford University Press.

Campbell, J. 1988. *The Power of Myth*. New York: Doubleday.

Campbell, J.M. 2008. "Pharmacy in Ancient Egypt." In R. David (ed.), *Egyptian Mummies and Modern Science*, 216–36. New York: Cambridge University Press.

Caplan, A.L. 1993. "The Concepts of Health. Illness and Disease." In W.F. Bynum and R. Porter (eds.), *Companion Encyclopedia of the History of Medicine*, 1: 233–48. London and New York: Routledge.

Carde, R. and Resh, C. 2012. *A World of Insects*. London: Harvard University Press.

Carlson, J.B. and Cherry, R. 1996. "Arthropods in Astronomy: Cases in 'Western' and Mesoamerican Enthoentomology." *American Entomologist* 42(3): 149–58.

Carter, D. 1992. *Butterflies and Moths*. New York: Dorling Kindersley.

Cashford, J. 2003. *The Moon: Myth and Image*. New York: Four Wall Eight Windows.

Cato, M.P. 1968. *On Agriculture*. Trans. W.D. Hooper. The Loeb Text. London: Heinnemann.

Celsus, A.C. 1935. *De Medicina*. Available online: https://penelope.uchicago.edu/Thayer/E/Roman/Texts/Celsus/3*.html (accessed February 14, 2022).

Celsus, A.C. 1935a. *De Medicina*. Available online: http://www.perseus.tufts.edu/hopper/text?doc=Cels.+6.7&fromdoc=Perseus%3Atext%3A1999.02.0142 (accessed February 14, 2022).

Chaddick, P.R. and and Leek, F.F. 1972. "Further Specimens of Stored Products Insects Found in Ancient Egyptian Tombs." *Journal of Stored Products Research* 8: 83–6.

Chamberlain, G. 2001. "Two Babies That Could Have Changed World History." *The Historian* 72.

Champollion Figeac, M. 1814. "Account of a New Species of Insect of the Genus *corynetes* of Fabritius, Observed in an Egyptian Mummy at Grenoble." *The Philosophical Magazine*, 1st Series 44: 131–3.

Chernetsov, V.N. 1963. "Concepts of the Soul Among the Ob Ugrians." In N. Michael (ed.), *Studies in Siberian shamanism*, 3–45. Toronto, Canada: University of Toronto Press.

Cherry, R. 1991. "Use of Insects by Australian Aborigines." *American Entomologist* 37: 8–13.

Cherry, R. 1993. "Insects in the Mythology of Native Americans." *American Entomologist* 39: 61–21.

Cherry, R. 1997. "Insect Names Derived From Greek and Roman Mythology." *American Entomologist* 43: 212–16.

Cherry, R. 2002. "The Functions of Insects in Mythology." *American Entomologist* 48: 134–6.

Cherry, R. 2004. "Praying Mantids as Symbols for Headhunting." *American Entomologist* 50(1): 12–16.

Cherry, R. 2005. "Magical Insects." *American Entomologist* 51(1): 11–16.

Cherry, R. 2011. "Insects and Death." *American Entomologist* 57(2): 82–5.

Cherry, R. 2012. "Insect Monsters in Mythology." *American Entomologist* 58(3): 138–40.

Cherry, R. 2015. "Insects and Divine Intervention." *American Entomologist* 61: 81–4.

Chugg, A.M. 2004. *The Lost Tomb of Alexander the Great*. London: Periplus Publishing London Ltd.

Cirillo, V.J. 2016. "'I Am the Baby Killer!' House Flies and the Spread of Polio." *American Entomologist* 62(2): 83–5.

Clausen, L. 1954. *Insect Fact and Folklore*. New York: Macmillan Co.

Clayton, P.A. 1994. *Chronicle of the Pharaohs*. London: Thames and Hudson.

Cloudsley-Thompson, J. 1976. *Insects and History*. New York: St. Martin's Press.

Cockburn, A. and Cockburn, E. 1981. "Autopsies in Mummies." *The New England Journal of Medicine: Boston* 305: 1534.

Cockburn, A., Cockburn, E., and Reyman, T.A., eds. 1998. *Mummies. Disease & Ancient Cultures*. Cambridge: Cambridge University Press.

Cohen, M.N. 1989. *Health and The Rise of Civilisation*. New Haven and London: Yale University Press.

Columella. 1845. *De Re Rustica*. Trans. L. du Bois. Paris: Panouche.

Conrad, L.I. 1995. "Epidemic Disease in Formal and Popular Thought in Early Islamic Society." In T. Ranger and P. Slack (eds.), *Epidemics and Ideas*, 77–99. Cambridge: Cambridge University Press.

Conroy, A.L. Bangirana, P., Muhindo, M.K., Kakuru, A., Jagannathan, P., Opoka, R.O., Liechty E.A., Nakalembe, M., Kamya, M.R., Dorsey, G., and John, C.C. 2019. "Case Report: Birth Outcome and Neurodevelopment in Placental Malaria Discordant Twins." *American Journal of Tropical Medicine and Hygiene* 100(3): 552–5.

Conti Bizarro, F. 2009. *Comici Entomologi*. Alessandria: Edizioni del Orso.

Cook, A.B. 1895. "The Bee in Greek Mythology." *Journal of Hellenistic Studies* 15: 1–24.

Corfield, J. 2012. "Health (1400–900)." In E. Ramsamy (ed.), *Cultural Sociology of the Middle East. Asia and Africa*, 2: 170–1. Africa: California: Sage Publications.

Corkhill, M. 2011. *Natural Remedies*. New York: Skyhorse Publishing.

Costantini, L., Tosi, M., and Vigna-Taglionti, A. 1977. "Typology and Socioeconomical Implications of Entomological Finds From Some Ancient Near Eastern Sites." *Paleorient* 3: 247–58.

Crane, E. 1978. "Places to Visit: The British Museum's Beeswax Treasures." *Bee World* 59(1): 39–40.

Crane, E. 1983. *The Archaeology of Beekeeping*. Ithaca, NY: Cornell University Press.

Crane, E. 1999. *The World History of Beekeeping and Honey Hunting*. New York: Routledge.

Crawford, P.L. 2007. "Beyond Paper: Use of Plants of the Cyperaceae Family in Ancient Egypt." In R. Cappers (ed.), *Fields of Change: Progress in African Archaeology*, 105–13. Groningen, The Netherlands: Groningen Institute of Archaeology.

Cunha, Burke A. 2004. "The Death of Alexander, the Great: Malaria or Typhoid Fever?" *Infectious Disease Clinics of North America* 18: 53–63.

Cunha, C.B. and Cunha, B.A. 2008. "Brief History of the Clinical Diagnosis of Malaria: From Hippocrates to Osler." *Journal of Vector Borne Diseases* 45: 194–9.

Curran, C.H. 1937. "Insect Lore of the Aztecs Revealing Early Acquaintance With Many of our Pests." *Journal of Natural History* 39: 196–203.

Curry, A. 1979. “The Insects Associated with the Manchester Mummies.” In A.R. David (ed.), *The Manchester Mummy Project*, 113–17. Manchester: Manchester Museum.

Curtis, C.F., Lines, J.D., Lu, B.L. and Renz, A. 1990. “Natural and Synthetic Repellents.” In C.F. Curtis (ed.), *Appropriate Technology in Vector Control*, 5–41. Boca Raton, FL: CRC Press. DOI: 10.1201/9781351069823.

Curtis, C.F., Lines, J.D., Carnevale, P., Robert, V., Boudin, C., Halna, J-M., Parzat, L., Gavin, P., Richard, A., Mouchet, J., Charlwood, J.D., Groves, P.M., Hossain, M.I., Kurihara, T., Ichimori, K., Zuzi, Li., Baolin, L., Majori, G., Sabitinelli, G., Coluzzi, M., Njunwa, K.J., Wilkes, T.J., Snow, R.W., and Lindsay, S.W. 2018. “Impregnated Bed Nets and Curtains Against Malaria Mosquitoes.” In C. Curtis (ed.), *Appropriate Technology in Vector Control*. Boca Raton. FL: CRC Press Ltd.

Curto, S. 1971. *The Military Art of the Ancient Egyptians*. Turin: Edizioni D’Arte Frateli Pozzo.

Dacke, M., Baird, E., Byrne, M., Scholtz, C.H., and Warrant, E.J. 2013. “Dung Beetles Use the Milky Way for Orientation.” *Current Biology* 23: 298–300.

David, R. 2002. *Religion and Magic in Ancient Egypt*. London: Penguin Books.

Davies, M. and Kathirithamby, J. 1986. *Greek Insects*. London: Duckworth.

Delobel, B. and Delobel, A. 2006. “Dietary Specialization in European Species Groups of Seed Beetles (Coleoptera: Bruchidae: Bruchinae).” *Oecologia* 149: 428–43.

Derry, D.E. 2001. “Appendix I: Report Upon the Examination of Tutankhamun Mummy.” In H. Carter, *The Tomb of Tutankhamen*. vol. II. London: Duckworth.

Derry, M.B. 1927. “Appendix I: Report Upon the Examination of Tut-ankh-Amen’s Mummy.” In H. Carter, *The Tomb of Tutankhamun*, vol. 2, *The Burial Chamber*, 101–13. London.

Dickson, J.H., Oeggl, K., and Handley, L.L. 2005. “The Iceman Reconsidered.” Available online: https://www.scientificamerican.com/article/the-iceman-reconsidered-2005-01/ (accessed February 15, 2022).

Dickson, P. 2006. *Dictionary of Middle Egyptian*. Available online: https://archive.org/details/DictionaryOfMiddleEgyptian/page/n1/mode/2up (accessed February 12, 2022).

Dillon, D.S. and Dillon, L.S. 1972. *A Manual of Common Beetles of Eastern North America*. 2 vols. New York: Dover Publications.

Diodorus Siculus. 1933. *The Library of History Books* I–II. Trans. C.H. Oldfather. Cambridge, MA: Harvard University Press.

Diodorus of Sicily, with an English trans. by C. Bradford Welles. 1963. London: William Heinemann Ltd.

Dobson, M. 2007. “*Murderous Contagion* (A Human History of Disease).” London: Quercus.

Donkin, R.A. 1977. “The Insect Dyes of Western and West-Central Asia.” *Anthropos* 72: 847–80.

Doumas, C. 1983. *Thera Pompeii of the Ancient Aegean*. London: Thames and Hudson.

Dow, R.P. 1917. “Studies in the Old Testament.” *Bulletin of the Brooklyn Entomological Society* 12: 64–9.

Dow, R.P. 1918. “Grasshoppers in the Old Testament.” *Bulletin of the Brooklyn Entomological Society* 13: 23–30.

Dunn, F.L. 1993. “Malaria.” In K.F. Kiple, *The Cambridge World History of Human Disease,* 855–62. Cambridge: Cambridge University Press.

Ebbell, B. 1937. *The Papyrus Ebers*. Copenhagen: Levin & Muksgaard.

Ebeid, N.I. 1999. *Egyptian Medicine in the Days of the Pharaohs*. Cairo?: General Egyptian Book Organisation.

Egan, R.B. 1988. "Two Complementary Epigrams of Meleager: *A.P.* vii 195 and 196." *Journal of Hellenic Studies* 108: 24–32.

Egan, R.B. 1994. "Cicadas in Ancient Greece: Ventures in Classical Tettigology." *Cultural Entomology Digest* 3: 20–5.

Egan, R.B. 1998. "Archias, Meleager, Tymnes: Dead Birds in Context." *Rheinisches Museum für Philologie* 141: 24–30.

Egan, R.B. 2004. "Eros, Eloquence and Entomo-psychology in Plato's *Phaedrus*." In R.B. Egan and M.A. Joyal (eds.), *Daimonopylai: Essays in Classics and the Classical Tradition Presented to Edmund G. Berry*, 65–88. Winnipeg: University of Manitoba Centre for Hellenic Civilization.

Egan, R.B. 2014. "Insects." In G.L. Campbell (ed.), *The Oxford Handbook of Animals in Classical Thought and Life*, 180–91. Oxford: Oxford University Press.

Ellis, L., Wahab, E.A., and Ratnasingan, M. 2013. "Religiosity and Fear of Death: A Three-Nation Comparison." *Mental Health, Religion and Culture* 16(2): 179–99.

Enterline, L. 2000. *The Rhetoric of the Body from Ovid to Shakespeare*. Cambridge: Cambridge University Press.

Erdman, A. 1971. *Life in Ancient Egypt*. New York: Dover Publications.

Erdoes, R. and Ortiz, A., eds. 1984. *American Indian Myths and Legends*. New York: Pantheon Books.

Essig, E.O. 1931. *A History of Entomology*. New York: Macmillan.

Estes, J.W. 1993. *The Medical Skills of Ancient Egypt*. Canton, MA: Science History Publications.

Evans, A. 1925. *The Ring of Nestor. A Glimpse into the Minoan After-world and a Sepulchral Treasure of Gold Signet-rings and Bead-seals from Thisbe, Boetia*. London: Macmillan and Co.

Evans, L. and Weinstein, P. 2021. "'Begone from Me, O Crooked-Lips!' Integrated Pest Management in Ancient Egypt." *American Entomologist* 67: 46–53.

Exner, S., Bogusch, G., and Sokiranski, R. 2004. "Cribra Orbitalia Visualized in Computed Tomography," *Annals of Anatomy* 186: 169–72.

Fagerlie, J. 1964. "Monies of Antiquity." *Natural History* 73: 20–5.

Fairhead, J. and Leach, M. 2003. "Termites, Society and Ecology: Perspectives from West Africa." In E. Motte-Florac and J.M.C. Thomas (eds.), *Les "insectes" dans la tradition orale ("insects" in oral literature and traditions)*, 197–219. Paris: Peeters.

Falco, R. 2017. *Cultural Genealogy: An Essay on Early Modern Myth*. London: Routledge.

Farrell, W.J. 1985. "The Role of Mandeville's Bee Analogy in 'The Grumbling Hive'." *Studies in English Literature 1500–1900* 25(3): 511–27.

Faulkner, R.O. 1994. *Egyptian Book of the Dead: Book of Going Forth by Day*. Chronicle Books.

Fenyo, M.D. 2012. "Malaria and Europeans (1400–900)." In E. Ramsamy (ed.), *Cultural Sociology of the Middle East. Asia and Africa*, vol. 2: *Africa* vol. 2: 170–1. California: Sage Publications.

Fewkes, J.W. 1910. "The Butterfly in Hopi Myth and Ritual." *American Anthropologist*, New Series 12: 576–94.

Filer, J.M. 1992. "Head Injuries in Egypt and Nubia: A Comparison of Skulls From Giza and Kerma." *Journal of Egyptian Archaeology* 78: 281–5.

Filer, J.M. 1995. *The Egyptian Bookshelf: Disease*. London: British Museum Press.

Filer, J.M. 1997. "Ancient Egypt and Nubia as a Source of Information for Cranial Injuries." In J. Carman (ed.), *Material Harm: Archaeological Studies of War and Violence*, 47–74. Glasgow: Cruithne Press.
Filer, J.M. 1998. "Revealing the Face of Artemidorus." *Minerva* 9(4): 21–4.
Filer, J.M. 2001. "Hygiene." In D. Redford (ed.), *The Oxford Encyclopedia of Ancient Egypt*, vol. 2: 133–6. Oxford: Oxford University Press.
Fogliazza, D.D. and Pagani, M. 1993. "Insect Pests in Stored Foodstuffs in Italy. Part 1: Coleoptera." *Tecnica Molitoria* 44: 937–51.
Forbes, R.J. 1956. *Studies in Ancient Technology*. Vol. IV. Leiden: Brill.
Forbes, R.J. 1957. *Studies in Ancient Technology*. Vol. V. Leiden: Brill.
Forbes, R.J. 1958. *Studies in Ancient Technology*. Vol. VI. Leiden: Brill.
Forbes, R.J. 1965. *Studies in Ancient Technology*. Vol. III. Leiden: Brill.
Forum Ancient Coins. Available online: https://www.forumancientcoins.com/catalog/roman-and-greek-coins.asp?pos=0&sold=1&vpar=1470&iop=50&galview=0&vorderby=pricedesc (accessed February 20, 2022).
Francis, J.E. and Kouremenos, A. 2016. *Apiculture in Roman Crete: New Perspectives*. Oxford: Oxbow Books.
Fraser, H.M. 1931. *Beekeeping in Antiquity*. London: University of London Press.
Frazier, J.G. 1930. *Myths of the Origin of Fire*. New York: Macmillan.
Gaudet, J. 2014. *Papyrus*. New York: Pegasus Books.
Genduso, P. 1960. "Observations on the Bruchids of Lentil (Lens esculenta Moench). 1st note. *Bruchus lentis* Fröel.-*Bruchus ervi* Fröel." *Bollettino dell'Istituto di Entomologia Agraria e dell' Osservatorio di Fitopatologia di Palermo* 10: 139–62.
George, A. 2003. *The Epic of Gilgamesh*. London: Penguin Books.
Ghaliounghui, P. 1963. *Magic and Medical Science in Ancient Egypt*. London: Hodder and Stoughton.
Gimbutas, M. 1974. *The Gods and Goddesses of Old Europe*. Berkeley: University of California Press.
Girling, M.A. 1984. "Investigations of a Second Insect Assemblage From the Sweet Track." *Somerset Levels Papers* 10: 79–91.
Giusti, E. 2014. "Virgil's Carthaginians at *Aen*. 1.430–6: Cyclopes in Bees' Clothing." *Cambridge Classical Journal* 60: 37–58.
Godolphin, F.R.B., ed. 1964. *Great Classical Myths*. New York: The Modern Library.
Goodall, J. 2022. Chimpanzee tool use. Available online: https://www.janegoodall.org.uk/chimpanzees/chimpanzee-central/15-chimpanzees/chimpanzee-central/19-toolmaking (accessed February 12, 2022).
Granville, A. 1825. "An Essay on Egyptian Mummies: With Observations on the Art of Embalming Among the Ancient Egyptians." *The Transactions of the Philosophical Society*, April 14: 269–316.
Griffith, F.L. and Thompson, H., eds. 1904. *The Demotic Magical Papyrus of London and Leiden*. London: Grevel & Co. (See also as Griffith, F.L. and Thompson, H., ed. 1974. *The Leyden Papyrus. An Egyptian Magical Book*. New York: Dover Publications, Inc.).
Grimal, P. 1965. "Greece: Myth and Logic." In P. Grimal (ed.), *Larousse World Mythology*, 97–176. New York: Prometheus Press.
Grimaldi, D. and M.S. Engel. 2005. *Evolution of the Insects*. Cambridge: Cambridge University Press.
Grinnell, G.B. 1899. "The Butterfly and Spider Among the Blackfeet." *American Anthropologist*, New Series 1: 194–6.

Gunn, B. 1916. "The Religion of the Poor in Ancient Egypt." *Journal of Egyptian Archaeology* 3: 81–94.

Guthrie, R.D. 2005. *The Nature of Paleolithic Art*. Chicago: University of Chicago Press.

Haffejee, I.E. and Moosa, A. 1985. "Honey in the Treatment of Infantile Gastroenteritis." *British Medical Journal* 290: 1866–87.

Halioua, B. and Ziskind, B. 2005. *Medicine in the Days of the Pharaohs*. Cambridge, Massachusetts, and London: The Belknap Press of Harvard University Press.

Halstead, D.G.H. 1993. "Keys for the Identification of Beetles Associated With Stored Products – II. Laemophloeidae, Passandridae and Silvanidae." *Journal of Stored Product Research* 29: 99–197.

Hamilton, E. 1942. *Mythology*. Boston: Little, Brown and Company.

Hannaway, C. 2001. "Environment and Miasmata." In W.F. Bynum and R. Porter (eds.), *Companion Encyclopedia of the History of Medicine*, vol 1: 292–307. London and New York: Routledge.

Hansen, N.B. 2001. "Insects." In D.B. Redford, *The Oxford Encyclopedia of Ancient Egypt*, 161–3. New York: Oxford University Press.

Hanski, I. 1992. "Are the Pyramids Deified Dung Pats?" In J. Adams (ed.), *Insect Potpourri: Adventures in Entomology*, 278–80. Gainesville, FL: Sandhill Crane Press.

Harer, W.B. and Tapp, E. 2014. "Augustus Bozzi Granville (1783–1872): Pioneer Obstetrician and Gynaecological Surgeon Who Performed the First Scientific Autopsy of a Mummy." *Journal of Medical Biography* 22(2): 101–7. DOI: 10.1177/0967772013480607.

Harpaz, I. 1973. "Early Entomology in the Middle East." In R.F. Smith, T.E. Mittler, and C.N. Smith (eds.), *History of Entomology*. Palo Alto, CA: Annual Reviews.

Harris, J.P. 1989. *Insect Imagery in Aristophanes*. MA thesis, University of Manitoba.

Harris, J.P. 2018. "Flies, Wasps, and Gadflies: The Role of Insect Similes in Homer, Aristophanes and Plato." *Mouseion* 15: 475–500.

Harrison, G. 1978. *Mosquitoes, Malaria and Man: A History of the Hostilities Since 1880*. London: John Murray.

Harrison, R.G. and Abdalla, A.B. 1972. "The Remains of Tutankhamun." *Antiquity* 46: 8–14.

Harrison, R.G., Connolly, R.C., Ahmed, S., Abdalla, A.B., and El Ghawaby, M. 1979. "A Mummified Foetus From the Tomb of Tutankhamun." *Antiquity* 53: 19–21.

Hawass, Z. and Saleem, S.N. 2011. "Mummified Daughters of King Tutankhamun: Archeologic and CT Studies." *American Journal of Roentology* 197: W829–W836.

Hawass, Z., Gad, Y.Z., Ismail, S., Khairat, R., Fathalla, D., Hasan, N., Ahmed, A., Elleithy, H., Ball, M., Gaballah, F., Wasef, S., Fateen, M., Amer, H., Gostner, P., Selim, A., Zink, A., and Pusch, C.M. 2010. "Ancestry and Pathology in King Tutankhamun's Family." *Journal of the American Medical Association* 303: 638–47.

Hellier, C.A. and Connolly, R.A. 2009. "A Re-assessment of the Larger Fetus Found in Tutankhamen's Tomb." *Antiquity* 83: 165–17.

Hempel, S. 2018. *The Atlas of Disease*. London: White Lion Publishing.

Hepburn, H.R. 1986. *Honeybees and Wax*. Berlin: Springer-Verlag.

Herodotus. 1920. *The Persian Wars*. Available online: https://www.loebclassics.com/view/herodotus-persian_wars/1920/pb_LCL117.319.xml (accessed April 19, 2023).

Herodotus. 1988. *The Histories*. Trans. A. de Sélincourt. London: Penguin Books.

Herodotus. 2013. *An Account of Egypt*. Available online: https://www.gutenberg.org/files/2131/2131-h/2131-h.htm (accessed February 12, 2022).
Hill, G.F. 1922. "Ancient Methods of Coining." *The Numismatic Chronicle and Journal of the Royal Numismatic Society*, Fifth Series 2: 1–41.
Hillson, S. 1980. "Chronic Anaemias in the Nile Valley." *Museum Applied Science Center for Archaeology* (MASCA) 1(6) (Mummy Supplement).
Hinton, H.E. 1945. *A Monograph of the Beetles Associated with Stored Products, I*. London: British Museum (N.H.).
Hinton, H.E. 1948. "A Synopsis of the Genus Tribolium Mac Leay with Some Remarks on the Evolution of the Species-Groups (Coleoptera, Tenebrionidae)." *Bulletin of Entomological Research* 39: 13–55.
Hitchcock, S.W. 1962. "Insects and Indians of the Americas." *Bulletin of the Brooklyn Entomological Society* 8: 181–7.
Hoffmann, A. 1954. "Coleoptérès Curculionides 2." *Faune de France* 59: 487–1208.
Hogue, C.L. 1975. "The Insect in Human Symbolism." *Terra* 13(3): 3–9.
Hogue, C.L. 1983. "Commentaries in Cultural Entomology. 3. An Entomological Explanation of Ezekial's Wheels." *Entomological News* 94(3): 73–80.
Hogue, C.L. 1987. "Cultural Entomology." *Annual Review of Entomology* 32: 181–99.
Holldobler, B. and E.O. Wilson. 1990. *The Ants*. Cambridge, MA: Belknap.
Hope, F.W. 1834. "Footnote." In T.J. Pettigrew (ed.), *History of Egyptian Mummies and an Account of the Worship and Embalming of the Sacred Animals*, 224. London: Longman, Rees, Orme, Brown, Green & Longman.
Hope, F.W. 1836. "Notice of Several Species of Insects Found in the Heads of Egyptian Mummies." *Transactions of the Entomological Society of London* 1: 11–13.
Hope, F.W. 1842. "Observations on Some Mummied Beetles Taken From the Inside of a Mummied Ibis." *Transactions of the Royal Entomological Society of London* 3: 191–3.
Horne, C. 1908. *The Bible and its Story*. New York: F.R. Niglutsch.
Horsfall, N. 2010. "Bees in Elysium." *Vergilius* 56: 39–45.
Hublin, J-J., Ben-Ncer, A., Bailey, S.E., Freidline, S.E., Neubauer, S., Skinner, M.M., Bergmann, I., Le Cabec, A., Benazzi, S., Harvati, K., and Gunz, P. 2017. "New Fossils From Jebel Irhoud, Morocco and the Pan-African Origin of Homo Sapiens." *Nature* 546(7657): 289. DOI: 10.1038/nature22336.
Huchet, J-B. 2010. "Archaeoentomological Study of the Insect Remains Found Within the Mummy of *Namenkhet Amun* (San Lazzaro Armenian Monastery, Venice/Italy)." *Advances in Egyptology* 1: 59–80.
Huchet, J-B. and Greenberg, B. 2010. "Flies, Mochicas and Burial Practices: A Case Study From Huaca de la Luna, Peru." *Journal of Archaeological Science* 37: 2846–56.
Huddilston, J.H. 1902. *Lessons From Greek Pottery*. New York: The Macmillan Company.
Hughes, J.D. 2014. *Environmental Problems of the Greeks and Romans: Ecology in the Ancient Mediterranean*. Baltimore, MD: Johns Hopkins University Press.
Hume, J., Hume, C., Lyons, E.J., and Day, K.P. 2003. "Malaria in Antiquity: A Genetics Perspective." *World Archaeology*. Available online: https://doi.org/10.1080/00438240320001113362 (accessed April 19, 2023).
Humphreys, M. 1997. "Malaria: 'Evil Air' and Mosquitoes." In K.F. Kiple (ed.), *Plague, Pox and Pestilence*, 98–103. London: Weidenfeld and Nicolson.

Hunt, L.B. 1980. "The Long History of Lost Wax Casting: Over Five Thousand Years of Art and Craftsmanship." *Gold Bulletin* 13(2): 63–79.

Icard-Gianolio, N. 1994. "Psyché-papillon." In *Lexicon Iconographicum Mythologiae Classicae*, VII. 1. Zurich/Munich: Artemis Verlag.

Ikram, S. and Dodson, A. 1998. *The Mummy in Ancient Egypt*. London: Thames and Hudson.

Ions, V. 1999. *The History of Mythology*. Surrey, England: Quadrillion Publishing Limited.

Jablonski, N.G. 2010. "The Naked Truth." *Scientific American* 302: 42–9.

Jackson, F.L.C. 2000. "Human Adaptations to Infectious Disease." In S. Stinson, B. Bogin, R. Huss-Ashmore, and D. O'Rourke (eds.), *Human Biology: An Evolutionary and Biocultural Perspective*, 273–94. New York: Wiley-Liss.

Jaeger, E.C. 1978. *A Source Book of Biological Names and Terms*, 3rd edn. Springfield, IL: Charles Thomas.

James, T.H.G. 1973. "Egypt: From the Expulsion of the Hyksos to Amenophis I." In I.E.S. Edwards et al. (eds.), *Cambridge Ancient History*, vol. 2, Part 1, 3rd edn, 289–312. Cambridge: Cambridge University Press.

Janko, R. 2017. "Tithonus, Eos and the Cicada in the Homeric Hymn to Aphrodite and Sappho." In C. Tsagali and S. Markantonatos (eds.), *The Winnowing Oar – New Perspectives in Homeric Studies*. Berlin, Germany: De Gruyter.

Jeziorska, M. 2008. "Palaeopathology at the Beginning of the New Millennium: A Review of the Literature." In D. Rosalie (ed.), *Egyptian Mummies and Modern Science*, 83–98. New York: Cambridge University Press.

Jobes, G. 1962. *Dictionary of Mythology, Folklore and Symbols*, vol. 1. New York: Scarecrow Press.

Jones, G. 1983. "The LM II Plant Remains." In M.R. Popham (ed.), *The Minoan Unexplored Mansion at Knossos*, 303–6. British School at Athens. London: Thames & Hudson.

Jones, J.E. 1962. "Hives and Honey of Hymettus: Beekeeping in Ancient Greece." *Archaeology* 29: 80–92.

Jones, R. 2012. *Mosquito*. London: Reaktion Books Ltd.

Jones, R. 2015. *House Guests. House Pests*. London: Bloomsbury (Natural History).

Jones, W.H.S. 1868. *Hippocrates Collected Works I. Hippocrates*. Cambridge, MA: Harvard University Press. Available online: http://www.perseus.tufts.edu/hopper/text?doc=Perseus:text:1999.01.0251:text=Epid.:book=1:chapter=2&highlight=malaria (accessed April 19, 2023).

Jones, W.H.S. 1907. *Malaria*. London: Macmillan and Company.

Jones, W.H.S. 1909. *Malaria and Greek History*. Manchester: Manchester University Publications.

Jordan, P. 2003. *Material Culture and Sacred Landscape: The Anthropology of the Siberian Khanty*. Walnut Creek, CA: Rowan and Littlefield Publishers.

Jørgensen, L.B. 2013. "The Question of Prehistoric Silks in Europe." *Antiquity* 87(336): 581–8.

Jung, C.G. 1964. "Man and His Symbols." Garden City, New York: Doubleday & Company Inc.

Kadar, Z. 1978. *Survivals of Greek Zoological Illuminations in Byzantine Manuscripts*. Budapest: Akadémiai Kiadó.

Kamal, H. 1975. *Encyclopaedia of Islamic Medicine*. Cairo?: General Egyptian Book Organisation.

Keimer, L. 1938. *Insectes de L'Egypte Ancienne*. Cairo.
Kenaway, M.A. and Abdel-Hamid, Y.M. 2015. "Insects in Ancient (Pharaonic) Egypt." *Egyptian Academic Journal of Biological Sciences: Entomology* 8(1): 15–32.
Kenward, H. 1999. "Pubic Lice (*Pthirus pubis* L.) Were Present in Roman and Medieval Britain." *Antiquity* 73(282): 911–15.
Khalifa, A. and Badawy, A. 1960. "The Systematics of *Trogoderma afrum* Priesner and *T. granarium* Everts (Coleoptera: Dermestidae)." *Proceedings of the Royal Entomological Society of London* 29(11–12): 145–52.
Kiple, K.F. 1993. "The Ecology of Disease." In W.F. Bynum and R. Porter (eds.), *Companion Encyclopedia of the History of Medicine*, vol. 2: 357–411. London and New York: Routledge.
Kislev, M. and Simchoni, O. 2007. "Hygiene and Insect Damage of Crops and Food at Masada." In J. Aviram, G. Foerster, E. Netzer, and G.D. Stiebel (eds.), *Masada VIII. The Yigael Yadin Excavations 1963–1965. Final Reports*, 133–70. Jerusalem, Israel Exploration Society & Hebrew University of Jerusalem.
Kislev, M.E., Hartmann, A., and Galili, E. 2004. "Archaeobotanical and Archaeoentomological Evidence from a Well at Atlit-Yam Indicates Colder, More Humid Climate on the Israeli Coast During the PPNC Period." *Journal of Archaeological Science* 31: 1301–10.
Klausnitzer, B. 1981. *Beetles*. New York: Exeter Books.
Klein, B.A. 2007. "Insects and Humans: A Relationship Recorded in Visual Art." In M. Bekoff (ed.), *Encyclopedia of Human-Animal Relationships* 92–9. Portsmouth, New Hampshire: Greenwood Publishers.
Koch, K. 1989. *Die Käfer Mitteleuropas. Ökologie*, 1. Krefeld: Goecke & Evers.
Koch, K. 1992. *Die Käfer Mitteleuropas. Ökologie 3*. Krefeld: Goecke & Evers.
Kritsky, G. 1993. "Beetle Gods, King Bees, & Other Insects of Ancient Egypt." *KMT* 4: 32–9.
Kritsky, G. 1997. "Insects and Other Arthropods of the Bible, the New Revised Version." *American Entomologist* 43: 183–8.
Kritsky, G. 2010. *The Quest for the Perfect Hive*. New York: Oxford University Press.
Kritsky, G. 2015. *The Tears of Re: Beekeeping in Ancient Egypt*. New York: Oxford University Press.
Kritsky, G. 2017. "Beekeeping From Antiquity Through the Middle Ages." *Annual Review of Entomology* 62: 249–64.
Kritsky, G. 2018. "The Quest for The Perfect Hive: Ancient Mediterranean Origins." In F. Hatjina, G. Mavrofridis, and R. Jones, eds., *Beekeeping in the Mediterranean from Antiquity to the Present*. Nea Moudania.
Kritsky, G. 2020. "Revisiting Prehistoric Honey Hunting at Bicorp, Spain." *American Bee Journal* 160: 1147–250.
Kritsky, G. and Cherry, R. 2000. *Insect Mythology*. San Jose: Writers Club Press.
Kritsky, G. and Smith, J. 2018. "Insect Biodiversity in Culture and Art." In P. Adler and R. Foottit (eds.), *Insect Biodiversity*, vol. 2: 869–8. New York: Wiley and Sons.
Lambert, W.G. 2013. *Babylonian Creation Myths*. Winona Lake, IN: Eisenbrauns.
Larew, H.G. 2002. "Insects: Evidence from Wall Paintings Sculpture, Mosaics, Carbonized Remains, and Ancient Authors." In W.F. Jashemski and F.G. Meyer, *The Natural History of Pompeii*, 315–46. Cambridge: University of Cambridge Press.
Larsen, C.S. 1999. *Bioarchaeology. Interpreting Behavior from the Human Skeleton*. Cambridge: Cambridge University Press.

Leake, C.D. 1952. *The Old Egyptian Medical Papyri*. Lawrence, KS: University of Kansas Press.

Leek, F.F. 1972. *The Human Remains from the Tomb of Tutankhamun*. Oxford: The Griffith Institute.

Leek, F.F. 1975. "Some Evidence of Bees and Honey in Ancient Egypt." *Journal of Egyptian Archaeology* 53: 51–8.

Levinson, H.Z. and Levinson, A.R. 1985. "Storage and Insect Species of Stored Grain and Tombs in Ancient Egypt." *Zeitschrift für angewandte Entomologie* 100: 321–39.

Levinson, H.Z. and Levinson, A.R. 1994. "Origin of Grain Storage and Insect Species Consuming Desiccated Food." *Anzeiger für Schädlingskunde, Pflanzenschutz, Umweltschutz* 67: 47–59.

Levinson, H.Z. and Levinson, A.R. 1996. "*Prionotheca coronata* Olivier (Pimeliinae, Tenebrionidae) Recognized as a New Species of Venerated Beetles in the Funerary Cult of Pre-Dynastic and Archaic Egypt 2." *Journal of Applied Entomology – Zeitschrift für angewandte Entomologie* 120: 577–85.

Levinson, H.Z. and Levinson, A.R. 2001. "Venerated Beetles and Their Cultural-Historical Background in Ancient Egypt." *Insekten als Symbole göttlicher Verehrung und Schädlinge des Menschen. Spixiana, Supplement* 27: 33–75.

Lichtheim, M. 1973. *Ancient Egyptian Literature*, vol. I: *The Old and Middle Kingdoms*. Berkeley: University of California Press.

Lieberman, D. 2013. *The Story of the Human Body*. Evolution. Health and Disease. London: Penguin Books.

Liebert, R.S. 2010. "Apian Imagery and the Critique of Poetic Sweetness in Plato's *Republic*." *Transactions of the American Philological Association* 140: 97–115.

Lloyd, A.B. 1976. *Herodotus Book II: Commentary 1–98*. Leiden: E.J. Brill.

Lloyd, G.E.R. (ed.), Chadwick, J., and Mann, W.N. (trans.). 1983. *Hippocratic Writings*. London: Penguin Books.

Locke, J. 2016. "Insect Remains." In D.J. Breeze, *Bearsden. A Roman Fort on the Antonine Wall*, 289–300. Edinburgh: Society of Antiquaries of Scotland.

Longrigg, J. 1995. "Epidemic Ideas and Classical Athenian Society." In T. Ranger and P. Slack (eds.), *Epidemics and* Ideas, 21–44. Cambridge: Cambridge University Press.

López, C., Saravia, C., Gomez, A., Hoebeke, J., and Patarroyo, M.A. 2010. "Mechanisma of Genetically-Based Resistence to Malaria." *Gene* 467: 1–12.

Lucas, A. and Harris, J.R. 2011. *Ancient Egyptian Materials and Industry*. Mineola, NY: Dover Publications.

Mackenzie, D. 1923. *Myths of Pre-Columbian America*. London: Gresham Publishing Co.

Macrone, M. 1992. *By Jove! Brush up Your Mythology*. New York: Cader.

Maegraith, B. 1948. *Pathological Processes in Malaria and Blackwater Fever*. Oxford: Blackwell Scientific Publications.

Majno, G. 1975. *The Healing Hand*. Cambridge: MA: Harvard University Press.

Manniche, L. 1989. *An Ancient Egyptian Herbal*. London: British Museum Publications Limited.

Manniche, L. 1999. *Sacred Luxuries: Fragrance, Aromatherapy, and Cosmetics in Ancient Egypt*. Ithaca, NY: Cornell University Press.

Margotta, R. 2001. *History of Medicine*. Great Britain: Chancellor Press.

Marshall, A. 2015. "Fly & Lion." *KMT* 26(3): 39–47.

Marshall, J.A., Haes, E., and Christopher, M. 1988. "Grasshoppers and Allied Insects of Great Britain and Ireland." Colchester: Harley Books.

Mason, Jr., H.M. 1974. *The Fantastic World of Ants*. New York: David McKay.

Mazar, A., Namdar, D., Panitz-Cohen, N., Neumann, R., and Weiner, S. 2008. "Iron Age Beehives at Tel Rehov in the Jordan Valley." *Antiquity* 82: 629–39.

Mazar, A. and Panitz-Cohen, N. 2007. "It is the Land of Honey: Beekeeping at Tel Rehov." *Near Eastern Archaeology* 70(4): 202–19.

McGavin, G.C. 2000. *Insects*. London: Dorling Kindersley.

McLaren, A. 1990. *A History of Contraception*. Oxford: Basil Blackwell Ltd.

McMichael, A.J. 2011. "Insights From Past Millennia Into Climatic Impacts on Human Health and Survival." *PNAS* 109(13): 4730–7.

Mercatante, A. 1978. *Good and Evil in Myth and Legend*. New York: Barnes and Noble.

Merrifield, M.P. 1846. *The Art of Fresco Painting*. London: Charles Gilpin and Arthur Wallis.

Metropolitan Museum of Art. 2017. "Terracotta Column-krater (Bowl for Mixing Wine and Water)." Available online: http://www.metmuseum.org/art/collection/search/254649 (accessed April 19, 2023).

Meyer, M., Arsuaga, J-L., de Filippo, C., Nagel, S., Aximu-Petri, A., Nickel, B., Martínez, I., Gracia, A., Bermúdez de Castro, J.M., Carbonell, E., Viola, B., Kelso, J., Prüfer, K., and Pääbo, S. 2016. "Nuclear DNA Sequences from the Middle Pleistocene Sima de los Huesos Hominins." *Nature*. DOI: 10.1038/nature17405.

Middleton, H.J. 1891. *The Engraved Gems of Classical Times*. Cambridge: Cambridge University Press.

Miller, J. 2008. "Dental Health and Disease in Ancient Egypt." In R. David (ed.), *Egyptian Mummies and Modern* Science, 52–70. New York: Cambridge University Press.

Miller, J.S. 1825. [No title]. *Transactions of the Linnaean Society of London* 14, 585.

Miller, R.L., Ikram, S., Armelagos, G.J., Walker, R., Jarer, W.B. Shiff, C.J., Baggett, D., Carrigan, M., and Maret, S.M. 1994. "Diagnosis of Plasmodium Falciparum Infections in Mummies Using the Rapid Manual ParaSight-F Test." *Transactions of the Royal Society of Tropical Medicine and Hygiene* 88(1): 31–2.

Mills, S.Y. 1993. *The Essential Book of Herbal Medicine*. London: Penguin Arkana.

Mitchell, R.T. and Zim, H.S. 1962. *Butterflies and Moths*. Racine, WI: Western.

Mitchell, S. 2004. *Gilgamesh: A New English Version*. New York: Free Press.

Monserrat, D. 1996. *Sex and Society in Graeco-Roman Egypt*. London and New York: Kegan Paul International.

Montell, C. and Zwiebel, L.J. 2016. "Mosquito Sensory Systems." In A.S. Raijhel (ed.), *Advances in Insect Physiology: Progress in Mosquito Research*, vol. 51. London: Academic Press.

Montgomery, B.E. 1959. "Arthropods and Ancient Man." *Bulletin de la Société Royale Entomologique d' Egypte* 5: 68–70.

Morge, G. 1973. "Entomology in the Western World in Antiquity and in Medieval times." In R. Smith, T. Mittler, and C. Smith (eds.), *History of Entomology*, 37–80. Palo Alto, CA: Annual Reviews, Inc.

Morkot, R.G. 2003. *Historical Dictionary of Ancient Egyptian Warfare*. Lanham, MA and Oxford: The Scarecrow Press.

Moss, W.J., Shah, S.N., and Morrow, R.H. 2017. "The History of Malaria and its Control." *International Encyclopedia of Public Health*. Available online: https://doi.org/10.1016/B978-012373960-5.00374-9 (accessed April 19, 2023).

Motte-Florac, E. and Thomas, J.M.C., eds. 2003. *Les "insectes" dans la tradition orale ("Insects" in Oral Literature and Traditions)*. Paris: Peeters.

Mouffet, T. 1658. *The Theater of Insects or Lesser Living Creatures*. London.

Mumcuoglu, Y.K. and Zias, J. 1988. "Head Lice, *Pediculus humanus capitis* (Anoplura: Pediculidae) from Hair Combs Excavated in Israel and Dated from the First Century B.C. to the Eighth Century A.D." *Journal of Medical Entomology* 25(6): 545–7.

Munro, J.W. 1966. *Pests of Stored Products*. London: Hutchinson.

Murnane, J.K. 2001. In D.B. Redford, *The Oxford Encyclopedia of Ancient Egypt*, 2: 219–21. New York: Oxford University Press.

Murray, M.A. 2004. *The Splendor That Was Egypt*. Mineola, NY: Dover Publications.

Nerlich, A. 2008. "Mummies May Have Had a Real Scare – Malaria." Available online: www.nbcnews.com/id/wbna27345796 (accessed April 19, 2023).

Newberry, P.E. et al. 1893. *Beni Hasan*, vol. 2. London: The Egypt Exploration Fund.

Newberry, P.E. 1908. *Scarabs*. London: Archibald Constable and Co., Ltd.

Nicholson, I. 1985. *Mexican and Central American Mythology*. New York: Peter Bedrick Books.

Nicholson, P.T. 2001. "Faience." In D.B. Redford, *The Oxford Encyclopedia of Ancient Egypt*, 2: 491–6. New York: Oxford University Press.

Nunn, J. 1996. *Ancient Egyptian Medicine*. London: British Museum Press.

Nunn, J.F. and Tapp, E. 2000. "Tropical Diseases in Ancient Egypt." *Transactions of the Royal Society of Tropical Medicine and Hygiene* 94: 147–53.

OED Online. December 2021. "mosquito, n." Oxford University Press. Available online: https://www-oed-com.uc.idm.oclc.org/view/Entry/122564?redirectedFrom=mosquito (accessed February 6, 2022).

Oldach, D.W., Richard, R.E., Borza, E.N., and Benitez, R.M. 1998. "A Mysterious Death." *The New England Journal of Medicine* 338: 1764–9.

Osten Sacken, C.R. 1893. "On the So-Called *Bugonia* of the Ancients and its Relation to Eristalis Tenax, a Two-Winged Insect." *Bulletino della Societa Entomological Italiana* 25: 186–217.

O'Sullivan, L., Jardine, A., Cook, A., and Weinstein, P. 2008. "Deforestation, Mosquitoes, and Ancient Rome: Lessons for Today." *BioScience* 58(8): 756–60.

Palladius, R.T.A. 1773 and 1779. *On Husbandry*. Ed. N. Lodge and S.J. Herrtage. *The Early English Text Society*, 52, 72.

Panagiotakopulu, E. 2000. "Archaeology and Entomology in the Eastern Mediterranean. Research into the History of Insect Synanthropy in Greece and Egypt." *British Archaeological Reports 836*, International Series.

Panagiotakopulu, E. 2001. "Fleas from Pharaonic Amarna." *Antiquity* 75: 499–500.

Panagiotakopulu, E. and Buckland, P.C. 1991. "Insect Pests of Stored Products from Late Bronze Age, Santorini, Greece." *Journal of Stored Products Research* 21: 179–84.

Panagiotakopulu, E. and Buckland, P.C. 1999. "*Cimex Iectularius* L., The Common Bed Bug from Pharaonic Egypt." *Antiquity* 73: 908–11.

Panagiotakopulu, E. and Buckland, P. 2017. "A Thousand Bites – Insect Introductions and Late Holocene Environments." *Quaternary Science Reviews* 156: 23–35.

Panagiotakopulu, E. and Buckland, P. 2018. "Early Invaders – Farmers, the Granary Weevil and Other Uninvited Guests in the Neolithic." *Biological Invasions* 20 (1): 219–33.

Panagiotakopulu, E. and van der Veen, M. 1997. “Synanthropic Insect Faunas From Mons Claudianus, a Roman Quarry Site in the Eastern Desert, Egypt.” In A.C. Ashworth, P.C. Buckland, and J.P. Sadler, eds., *Studies in Quaternary Entomology—An Inordinate Fondness for Insects.* Quaternary Proceedings 5: 199–206.

Panagiotakopulu, E., Buckland, P.C., and Kemp, B. 2010. “Underneath Ra-Nefer’s House Floors: Archaeoentomological Investigations of an Elite Household in the Main City at Amarna, Egypt.” *Journal of Archaeological Science* 37: 474–81.

Panagiotakopulu, E., Buckland, P.C., Day, P., Sarpaki, A., and Doumas, C. 1995. “Natural Insecticides and Insect Repellents in Antiquity: A Review of the Evidence.” *Journal of Archaeological Science* 22: 705–10.

Panagiotakopulu, E., Buckland, P.C., Day, P.M., Doumas, C., Sarpaki, A., and Skidmore, P. 1997. “A Lepidopterous Cocoon from Thera and Evidence for Silk in the Aegean Bronze Age.” *Antiquity* 71: 420–9.

Panagiotakopulu, E., Higham, T., Sarpaki, S., Buckland, P.C., and Doumas, C. 2013. “Ancient Pests, the Season of the Santorini Minoan Volcanic Eruption and a Date from Insect Chitin.” *Naturwissenschaften* 100: 683–9.

Papagrigorakis, M.J., Yapijakis, C., and Synodinos, P.N. 2008. “Typhoid Fever Epidemic in Ancient Athens.” In D. Raoult and M. Drancourt (eds.), *Paleomicrobiology: Past Human Infections*, 161–73. Berlin and Heidelberg: Springer-Verlag.

Papagrigorakis, M.J., Yapijakis, C., Synodinos, P.N., and Baziotopoulou-Valavani, E. 2006. “DNA Examination of Ancient Dental Pulp Incriminates Typhoid Fever as a Probable Cause of the Plague of Athens.” *International Journal of Infectious Diseases* 10: 206–14.

Parker, N. 2013. “Angel in the Marble.” Available online: https://nilsaparker.medium.com/the-angel-in-the-marble-f7aa43f333dc (accessed April 19, 2023).

Parkinson, R.B. 1991. *Voices from Ancient Egypt*. London: British Museum Press.

Pataki, E. 2013. *Tettix: a tücök mint szakrális-poétikai szimbólum az ókori görög irodalomban*. Budapest: Szent István Társulat.

Pays, J-F. 2010a. “Plasmodium falciparum < toutankhamonensis >.” *Bulletin of the Société Pathologique Exotique* 103: 65–8.

Pays, J-F. 2010b. “Toutankhamon et sicklanémie.” *Bulletin of the Société Pathologique Exotique* 103: 346–7.

Peacock, E.R. 1993. *Adults and Larvae of Hide, Larder and Carpet Beetles and their Relatives (Coleoptera: Dermestidae) and of Derodontid Beetles (Coleoptera: Derodontidae).* Handbooks for the Identification of British Insects 5, 3. Royal Entomological Society of London.

Pemberton, J. 2015. *Myths and Legends – From Cherokee Dances to Voodoo Trances.* New York: Chartwell Books.

Pemberton, R.W. 2003. “Persistence and Change in Traditional Uses of Insects in Contemporary East Asian Cultures.” In E. Motte-Florac and J.M.C. Thomas (eds.), *Les “insectes” dans la tradition orale (“Insects” in Oral Literature and Traditions)*, 139–54. Paris: Peeters.

Petrie, W.M.F. 1898. *Deshasheh*. London: The Egypt Exploration Fund.

Pettigrew, T.J. 1834. *History of Egyptian Mummies and an Account of the Worship and Embalming of the Sacred Animals.* London: Longman, Rees, Orme, Brown, Green & Longman.

Pfeiffer, C.F. and Harrison, E.F., eds. 1962. *The Wycliffe Bible Commentary.* Nashville, TN: The Southwestern Company.

Pliny the Elder. 1855. *The Natural History of Pliny*. Available online: http://www.perseus.tufts.edu/hopper/text?doc=Perseus:abo:phi,0978,001:29 (accessed April 19, 2023).

Pliny the Elder. 1856. *The Natural History of Pliny*. London: Henry G. Bohn.

Plutarch. 1985. "Alexander [356–23 BC]" in *The Age of Alexander*. Trans. I. Scott-Kilvert. London: Penguin Books Ltd.

Porter, R. 2003. *Blood and Guts*. New York: W.W. Norton & Company.

Prete, F.R. and Wolfe, M.M. 1992. "Religious Supplicant, Seductive Cannibal, or Reflex Machine? In Search of the Praying Mantis." *Journal of the History of Biology* 25(1): 91–136.

Quick, J.D. (with B. Fryer). 2021. *The End of Epidemics*. London: Scribe Publications.

Quintus, C.R. and Yardley, J. trans. 1984. *The History of Alexander*. London: Penguin Books Ltd.

Quirke, S. and Spencer, J., eds. 1992. *The British Museum Book of Ancient Egypt*, London: British Museum Press.

Rageot, M., Peche-Quilichini, K., Py, V., Filippi, J.J., and Fernandez, X. 2015. "Exploitation of Beehive Products, Plant Exudates and Tars in Corsica During the Early Iron Age." *Archaeometry* 58: 315–32.

Ransome, H. 1986. *The Sacred Bee in Ancient Times and Folklore*. Bridgwater, England: Bee Books New and Old.

Ransome, H.M. 2004. *The Sacred Bee in Ancient Times and Folklore*. Mineoloa, NY: Dover Publications.

Ratcliffe, B.C. 1990. "The Significance of Scarab Beetles in the Enthnoentomology of Non-Industrial, Indigenous Peoples." In D.A. Posey, W.L. Overal, C.R. Clement, M.J. Plotkin, E. Elisabetsky, C. Novaes da Mota, and J.F. Pessôa de Barros (eds.), *Ethnobiology: Implications and Applications. Proceedings of the First International Congress of Ethnobiology*, 159–85. Para, Brazil: Museu Paraense Emílio Goeldi.

Rawlinson, G. 1859. *Herodotus, the Histories*. New York: D. Appleton Company.

Richter, G.M.A. 1924. *The Craft of Athenian Pottery; An Investigation of the Technique of Black-Figured and Re-Figured Athenian Vases*. New Haven, CT: Yale University Press.

Riegel, G.T. 1979. "The Fly Whisk." *Bulletin of the Entomological Society of America* 25(3): 196–9.

Riegel, G. 1994. "The Cicada in Chinese Folklore." *Cultural Entomology Digest* 3: 18–19.

Rittner, O. 2011. "A Revised Checklist of the Genus *Scarabaeus* Linnaeus 1758 (Coleoptera: Scarabaeidae) Occurring in Israel with a First Record of *Scarabaeus (Scarabaeus) irakensis* Stolfa." *Animma.X* 40: 1–4.

Roberts, C. and Keith, M. 1995. *The Archaeology of Disease*. New York: Cornell University Press.

Robins, G. 1993. *Women in Ancient Egypt*. London: British Museum Press.

Robins, G. 2001. "Grid systems." In D.B. Redford, *The Oxford Encyclopedia of Ancient Egypt*, 1: 68–71. New York: Oxford University Press.

Robins, G. 2008. *The Art of Ancient Egypt*. Cairo: The American University in Cairo Press.

Roffet-Salque, M., Regert, M., Evershed, R.P., Outram, A.K., Cramp, L.J.E., Decavallas, O., Dunne, J., Gerbault, P., Mileto, S., Mirabaud, S., Pääkkönen, M., Smyth, J., Šoberl, L., Whelton, H.L., Alday-Ruiz, A., Asplund, H., Bartkowiak,

M., Bayer-Niemeier, E., Belhouchet, L., Bernardini, F., Budja, M., Cooney, G., Cubas, M., Danaher, E.M., Diniz, M., Domboróczki, L., Fabbri, C., González-Urquijo, J.E., Guilaine, J., Hachi, S., Hartwell, B.N., Hofmann, D., Hohle, I., Ibáñez, J.J., Karul, N., Kherbouche, F., Kiely, J., Kotsakis, K., Lueth, F., Mallory, J.P., Manen, C., Marciniak, A., Maurice-Chabard, B., McGonigle, M.A., Mulazzani, S., Özdoğan, M., Perić, O.S., Perić, S.R., Petrasch, J., Pétrequin, A., Pétrequin, P., Poensgen, U., Pollard, C.J., Poplin, F., Radi, G., Stadler, P., Stäuble, H., Tasić, N., Urem-Kotsou, D., Vuković, J.B., Walsh, F., Whittle, A., Wolfram, S., Zapata-Peña, L., and Zoughlami, J. 2015. "Widespread Exploitation of the Honeybee by Early Neolithic Farmers." *Nature* 527: 226–31.

Roth, S., Balvín, O., Siva-Jothy, M.T., Di Iorio, O., Benda, P., Calva, O., Faundez, E.I., Khan, F.A.A., McFadzen, M., Lehnert, M.P., Naylor, R., Simov, N., Morrow, E.H., Willassen, E., and Reinhardt, K. 2019. "Evolved Before Their Bat Hosts and Did Not Co-speciate with Ancient Humans." *Current Biology*. DOI: 10.1016/j.cub.2019.04.048.

Royal Society for the Prevention of Cruelty to Animals (RSPCA). 2019. "Unsung Heroes." *RSPCA Animal Guardian* 11.

Rühli, F.J. and Ikram, S. 2014. "Purported Medical Diagnoses of Pharaoh Tutankhamun. c. 1325 BC." *HOMO – Journal of Comparative Human Biology* 65: 51–63.

Russman, E.R. 1990. *Egyptian Sculpture: Cairo and Luxor*. London: British Museum Publications Ltd.

Saggs, H.W.F. 1984. *The Might That Was Assyria*. London: Sidgwick & Jackson.

Sallares, R. 2002. *Malaria and Rome: A History of Malaria in Ancient Italy*. Oxford: Oxford University Press.

Sandison, A.T. and Tapp, E. 1998. "Disease in Ancient Egypt." In T.A. Cockburn, E. Cockburn, and T.A. Reyman (eds.), *Mummies. Disease & Ancient Cultures*, 38–58. Cambridge: Cambridge University Press.

Sarpaki, A.A. 1992. "The Palaeoethnobotany of the West House Akrotiri, Thera A Case Study." *Annual of the British School at Athens*. DOI: https://doi.org/10.1017/S0068245400015136.

Saunders, N.J. 2006. *Alexander's Tomb*. New York: Basic Books.

Sauvage, A. 1970. "Les Insectes dans la Poésie romaine." *Latomus* 29: 269–96.

Scarre, C. 1992. *Herodotu:s The Histories*. Trans. from the text of Baehr by H. Cary (ed.). London: The Folio Society.

Scheidel, W. 2001. *Death on the Nile: Disease and the Demography of Roman Egypt*. Leiden, Netherlands: Brill.

Schenk, E.T. and McMasters, J.H. 1956. *Procedure in Taxonomy*. Stanford, CA: Stanford University Press.

Schering Corporation. 1955. *Medicine and Pharmacy. An Informal History*, vol. I: *Ancient Egypt*. Bloomfield, NJ: Schering Corporation.

Schmidt, T. 1998. "Der Kornkäfer Sitophilus granarius Schön. (Curculionidae) aus der Schuttschicht des bankeramischen Brunnens von Erelenz-Kückhoven. Öst Rhein. Amt für Bodendenkmalpflege (Hg.) (Brunnen der Jungteinzeit. Internat. Symposium Erkelenz 27–20 Okt. 1997)." *Materialen zur Bodendenkmalpflege* 11: 261–9.

Sears, M. 2010. "Warrior Ants: Elite Troops in the *Iliad*." *Classical World* 103: 139–55.

Sharp, D. 1895. "Insects." *Cambridge Natural History*.

Shaw, I. and Boatwright, D. 2008. "Ancient Egyptian Warfare." In P. de Souza (ed.), *The Ancient World at War*. London: Thames and Hudson.

Sherman, I.W. 2007. *Twelve Diseases That Changed Our World*. Washington: DC: ASM Press.

Shilleto, A.R. 1886. *Pausanias' Description of Greece*, vol. 2. London: George Bell and Sons.

Sidwell, K. 2009. *Aristophanes the Democrat: the Politics of Satirical Comedy During the Peloponnesian War*. Cambridge: Cambridge University Press.

Simmonds, M., Howes, M-J., and Irving, J. 2016. *The Gardener's Companion to Medicinal Plants (Royal Botanic Gardens. Kew)*. London: Frances Lincoln Limited.

Smith, D.N. 2004. "The Insect Remains From the Well." In M.C. Bishop, ed., *Inveresk Gate: Excavations in the Roman Civil Settlement at Inveresk, East Lothian, 1996–000*. STAR Monograph, vol. 7, 81–8.

Smith, D.N. 2011. "Insects From Northfleet." In C. Barnett, J.J. McKinley, E. Stafford, J.M. Grimm, and C.J. Stevens, eds., *Settling the Ebbsfleet Valley: High Speed I Excavations at Springhead and Northfleet, Kent. The Late Iron Age, Roman and Medieval Landscape 3. Late Iron Age to Roman Human Remains and Environmental Reports, 88–90 +Table 75*, 243–62. Oxford and Salisbury: Oxford-Wessex Archaeology.

Smith, S. 2014. *Man and Animal in Severan Rome*. Cambridge: Cambridge University Press.

Smith, W.S. 1998. *The Art and Architecture of Ancient Egypt*. New Haven and London: Yale University Press.

Solomon, M.E. 1965. "Archaeological Records of Storage Pests: *Sitophilus granarius* (L.) (Coleoptera, Curculionidae) from an Egyptian Pyramid Tomb." *Journal of Stored Product Research* 1: 105–7.

Soren, D., Fenton, T., and Birkby, W. 1995. "The Late Roman Infant Cemetery near Lugnano in Teverina, Italuy: Some Implications." *Journal of Paleopathology* 7(1): 13–42.

Southwood, T.R.E. 1977. "Entomology and Mankind." *American Scientist* 65: 30–9.

Speck, F.G. 1935. *Naskapi: The Savage Hunters of the Labrador Peninsula*. Norman: University of Oklahoma Press.

Spence, L. 2005. *The Myths of Mexico and Peru*. New York: Barnes and Noble.

Staal, J.D.W. 1988. *The New Patterns in the Sky: Myths and Legends of the Stars*. Blacksburg, VA: McDonald and Woodward Publishing Company.

Starr, C.K. and Bozzoli de Willie, M.E. 1990. "Social Wasps Among the Bribri of Costa Rica." In D.A. Posey, W.L. Overal, C.R. Clement, M.J. Plotkin, E. Elisabetsky, C. Novaes da Mota, and J.F. Pessôa de Barros (eds.), *Ethnobiology: Implications and Applications*. Proceedings of the First International Congress of Ethnobiology (Belém, 1988). Museu Paraense Emílio Goeldi, Belém, Para, Brazil.

Stewart, A.K. 2012. "Malaria' in Edward Ramsamy (ed)." *Cultural Sociology of the Middle East. Asia and Africa, vol 2:* Africa, 318–20. California: Sage Publications. Inc.

Strabo. 1930. *Geography Book XVI*. Vol. VII. Trans. H.L. Jones. The Loeb Classical Library. London: Heineman Ltd.

Strabo. 1932. *Geography Book XVII*. Vol. VIII. Trans. H.L. Jones. The Loeb Classical Library. London: Heineman Ltd.

Strouhal, E. 1991. “Vertebral Tuberculosis in Ancient Egypt and Nubia.” In D.J. Ortner and A.C. Aufderheide (eds.), *Human Paleopathology*, 181–250. Washington and London: Smithsonian Institution Press.
Stuart-Macadam, P. 1991. “Porotic Hyperostosis: Changing Interpretations.” In D.J. Ortner and A.C. Aufderheide (eds.), *Human Paleopathology*, 40–51. Washington and London: Smithsonian Institution Press.
Swan, L.A. and Papp, C.S. 1972. *The Common Insects of North America*. New York: Harper & Row.
Taha, E.T., Gray, R.H., and Mohamedani, A.A. 1993. “Malaria and Low Birth Weight in Central Sudan.” *American Journal of Epidemiology* 138: 318–25.
Tapp, E. and Wildsmith, K. 1992. “The Autopsy and Endoscopy of the Leeds Mummy.” In A.R. David and E. Tapp (eds.), *The Mummy's Tale*, 132–53. London: Michael O'Mara.
Tedlock, D. 1985. *Popol Vuh*. New York: Simon and Schuster.
Theophrastus. 1916. *Enquiry into Plants I, II*. Trans. A. Hort. Loeb Text. London: Heinnemann.
Theophrastus. 1976. *De Causis Plantarum I, II*. Trans. B. Einarson and G.K.K. Link. Loeb Text. London: Heinnemann.
Theophrastus. 1990. *De Causis Plantarum III, IV*. Trans. B. Einarson and G.K.K. Link. Loeb Text. London: Heinemann.
Thorley, J. 1971. “The Silk Trade Between China and the Roman Empire at Its Height, “Circa” A.D. 90–130.” *Greece & Rome* 18(1): 71–80.
Thucydides. 2013. *History of the Peloponnesian War. Book 7*. Ed. W. Crawley. Available online: https://en.wikisource.org/wiki/History_of_the_Peloponnesian_War/Book_7 (accessed April 19, 2023).
Thucydides. 2013a. *History of the Peloponnesian War. Book 2*. Ed. W. Crawley. Available online: https://en.wikisource.org/wiki/History_of_the_Peloponnesian_War/Book_2 (accessed April 19, 2023).
Timmann, C. and Meyer, C.G. 2010. “Malaria, Mummies, Mutations: Tutankhamun's Archaeological Autopsy.” *Tropical Medicine and International Health* 15: 1278–80.
Torpey, A. 2016. “The Gardens of Ancient Egypt.” In H. Selin (ed.), *Encyclopaedia of the History of Science, Technology, and Medicine in Non-Western Cultures*, 1944–50. Berlin/Heidelberg: Germany Springer Science+Business Media Dordrech.
Valamoti, S.M. and Buckland, P.C. 1995. “An Early Find of *Oryzaephilus surinamensis* L. (Coleoptera: Silvanidae) from Final Neolithic Mandalo, Macedonia, Greece.” *Journal of Stored Product Research* 31: 307–9.
Van Buren, E.D. 1937–1939. “The Scorpion in Mesopotamian Art and Religion.” *Archiv für Orientforschung* 12: 1–28.
van der Toorn, K., Becking, B., and van der Horst, P.W., eds. 1999. *Dictionary of Deities and Demons in the Bible (DDD)*. Grand Rapids, MI: William B. Eerdmans Publishing Company.
Varro, M.T. 1934. *De Re Rustica*. Available online: http://penelope.uchicago.edu/Thayer/e/roman/texts/varro/de_re_rustica/1*.html (accessed April 19, 2023).
Varro, P.T. 1968. *On Agriculture*. Trans. W.D. Hooper. The Loeb Text. London: Heinnemann.
Verdet, J. 1987. *The Sky-Mystery, Magic and Myth*. New York: H. Abrams Inc.
Vilímková, M. 1969. *Egyptian Jewelry*. London: Paul Hamlyn.
Waldbauer, G. 2012. “People and Insect Plagues.” In R.T. Cardé and V.H. Resh (eds.), *A World of Insects*, 112–29. London: Harvard University Press.

Waterman, T.T. 1920. “Yurok Geography.” *University of California Publication: American Archaeology and Ethnology* 16: 177–314.

Wente, E. 1990. *Letters From Ancient Egypt*, vol. 1. Atlanta, GA: Scholars Press.

West, M.L. 1997. *The East Face of Helicon: West Asiatic Elements in Greek Poetry and Myth*. Oxford: Oxford University Press.

Wilkins, R. 1994. *(The Fireside Book of) Deadly Diseases*. London: Robert Hale.

Wilkinson, R.H. 2003. *The Complete Gods and Goddesses of Ancient Egypt*. London: Thames and Hudson.

Wilkinson, R.H. 2008. *Egyptian Scarabs*. Oxford: Shire Egyptology.

Wilkinson, T. 2005. *The Thames & Hudson Dictionary of Ancient Egypt*. London: Thames & Hudson.

Wilson, L.G. 1993. “Fevers.” In W.F. Bynum and R. Porter (eds.), *Companion Encyclopedia of the History of Medicine*, vol. 1: 408–37. London and New York: Routledge.

Wood, J.G. 1869. *Bible Animals: Being description of every living creature mentioned in the scriptures from the ape to coral*. London: Longmans, Green, Reader, and Dyer.

World Health Organization (WHO). 2002. Lymphaic filariasis. Available online: https://www.who.int/news-room/fact-sheets/detail/lymphatic-filariasis (accessed May 20, 2023).

Wright, B. 1977. *Hopi Kachinas*. Flagstaff, AZ: Northland Press.

Wyman, L. and Bailey, F. 1964. *Navaho Indian Ethnoentomology*. Albuquerque, NM: University of New Mexico Press.

Xenophon. 1918. *Hellenica*. Available online: http://www.perseus.tufts.edu/hopper/text?doc=Perseus:text:1999.01.0206:book=5:chapter=3&highlight=fever (accessed April 19, 2023).

Xenophon, B. 1918a. *Hellenica*. Ed. L. Carleton. Available online: http://www.perseus.tufts.edu/hopper/text?doc=Perseus%3Atext%3A1999.01.0204%3Abook%3D1%3Achapter%3D6%3Asection%3D16 (accessed February 12, 2022).

Yancey, P.H. 1952. *Origins from Mythology of Biological Names and Terms*. BIOS Classroom Series No. 5. Mt. Vernon, Iowa.

Zacher, F. 1934a. “Beiträge zur Geschichte, Verbreitung und Oekologie der Vorratsschädlinge.” *Entomologische Beihefte Berlin-Dahlem* 1: 83–6.

Zacher, F. 1934b. “Vorratsschädlinge und Speicherwirtschaft im alten und neuen Ägypten.” *Forschung und Fortschritte* 10: 347–8.

Zacher, F. 1937. “Vorratsschädlinge und Vorratsschutz, ihre Bedeutung für Volksernährung und Weltwirtschaft.” *Zeitschrift für hygienische Zoologie und Sc hädlingsekämpfung* 29: 193–202.

Ziskind, B. 2009. “Malaria: A Plague of Ancient Egypt” [in French]. *La Revue du Praticien* 59: 58–9.

Zumla, A. and Lulat, A. 1989. “Honey – A Remedy Rediscovered.” *Journal of the Royal Society of Medicine* 82(July): 384–5.

ACKNOWLEDGMENTS

Chapter 2

This chapter is dedicated to the memory of Silvian Ghost, much loved and much missed.

I would like to thank the editor Gene Kritsky (and Bloomsbury Publishing) for inviting me to contribute this chapter to this new series on Insects; Gene is also to be thanked further for his patience and advice concerning the text and illustrations and for drawing my attention to the article by Evans and Weinstein.

Richard Barritt is also thanked for help with preparing photographs and assisting with technological issues. Thanks are due to Annette Bugansky (Mary Ward Centre, London, England) for "emergency help" with research.

Grateful thanks are due to the following institutions for allowing me, over the years, to examine biological material in their keeping: the Duckworth Centre for Biological Research, Cambridge, England and the British Museum Department of Ancient Egypt and Sudan, London, England.

The British Library, London, is most sincerely thanked for research support. In particular thanks to all the Humanities Reading Room Staff, especially the Reference and Issues Desk Staff: in particular, Mel and Mukhtar provided invaluable and much-appreciated support and advice. In the same department grateful thanks are also due to James, Lena, Rachel, Nick, and Sandra (Head of Department). Also, in the Humanities Department thanks also to Mersina-Eirini Koutelou for discussing the Greek words for "gnat/s" and "mosquito/es."

INDEX